INDIGO DISTRICTS 1860

THE BLUE DEVIL:
Indigo and Colonial Bengal

THE BLUE DEVIL:
Indigo and Colonial Bengal

with an English translation of
Neel Darpan
by
Dinabandhu Mitra

AMIYA RAO AND B.G. RAO

DELHI
OXFORD UNIVERSITY PRESS
CALCUTTA BOMBAY MADRAS
1992

Oxford University Press, Walton Street, Oxford OX2 6DP

New York Toronto
Delhi Bombay Calcutta Madras Karachi
Kuala Lumpur Singapore Hong Kong Tokyo
Nairobi Dar es Salaam
Melbourne Auckland

and associates in
Berlin Ibadan

SBN 0 19 563090 4

Laserset at Bookshelf, New Delhi 110 003.
Printed at Rekha Printers, New Delhi 110 020
and published by S.K. Mookerjee, Oxford University Press
YMCA Library Building, Jai Singh Road, New Delhi 110 001.

For Aparna

Those who have been set on golden chairs to write
Will be questioned about those who
Wove their coats.
Not for their elevated thoughts
Will their books be scrutinised, but
Any casual phrase that suggests
Something about those who wove coats
Will be read with interest....

... at that time will be praised
Those who sat on the bare ground to write
Those who sat among the lowly
Those who sat with the fighters.

Bertolt Brecht

Contents

Preface and Acknowledgements

It is a sad story and a very personal one. But it has to be told at least to explain the incredible delay in completing this publication.

When we were both young, my husband and I had taken up the translation of *Neel Darpan*; we did the first two scenes and meant to complete the rest before long. But that was not to be. Soon came India's independence. New freedom brought euphoria and ecstacy, but also new problems, new demands on our time, new pressures on our minds. *Neel Darpan* was no longer the priority. Years passed as they always do. Much later, when we were old and thought that time was at last our own, we fished out our old manuscript, promising ourselves to complete the translation. But again, that was not to be; the promise could not be kept. My husband took seriously ill and worry had the priority over *Neel Darpan*. Time was not our own, as we had thought it would be; it became the master and took him. Again years passed; but I had to keep the promise.

Many friends have helped me in my work, and though I mention a few, I am beholden to all of them. I am grateful to Blair B. King, who sent me a copy of his *Blue Mutiny*, which first helped me in getting a perspective of the period I was writing about. Samprit Chatterjee found time to go through the translation and suggested several improvements. Geraldine Forbes went through the first draft of the introduction, made many valuable suggestions and helped in obtaining some of the illustrations. For the preparation of this book I offer my grateful thanks to my publisher.

But it is to the members of my family that I owe this book—particularly to my younger daughter.

Amiya Rao

Glossary

amin	a land surveyor
amla	an office clerk
anna	one-sixteenth of one rupee
banyan	a broker, a financier
baro babu	litt. 'big *babu*', (*babu* = gentleman), the eldest son of a family
baro thakur	husband's eldest brother
benamdar	person to whom property has been transferred only nominally to evade the law.
benami	nominal transfer of property
bibi	a non-Hindu woman, and especially an English woman
bigha	a measure of land (approximately half an acre)
bou / bouma	a wife, housewife, a daughter-in-law
Brindaban	a Hindu centre of pilgrimage in north India, famous for its temples of Krishna
chamar	a low Hindu caste the traditional occupation of whose members is tanning; figuratively a merciless fellow
char	land which rises above the water level after the river recedes
chhoto bou	litt. 'small *bou*' of *bou* above the younger daughter-in-law
crore	ten million rupees
cutcherry	court houses; a *zamindar's* court and officer; and the office of an indigo planter
dada	elder brother or cousin; any senior man not old enough to belong to the preceding generation

dewan	a steward in charge of an establishment
didi	elder sister, cousin or sister-in-law; any senior woman not old enough to belong to the preceding generation
Durga Puja	annual autumn festival at which the goddess Durga is worshipped
ghat	flight of steps descending to a river, a landing place
gomastha	agent, or secretary, who supervised the production of indigo
hakim	a magistrate, a judge
Hari	God
havildar	an Indian sergeant
ilaqa (be-ilaqa)	land that belonged (did not belong) to a planter
izarah	a lease of land
jotedar	a farmer
kabiraj	a physician who follows the traditional Hindu (*ayurvedic*) system
katha	narration of mythological stories
kaibarta / keaot	caste of fishermen
kavi-wala	professional narrator of stories
kayastha	a high Hindu caste
kulin	a family (or member of a family) recognized as belonging to a group of Brahmin families on whom special honour was bestowed in medieval Bengal by King Ballal Sen. A high sub caste of Brahmins in Bengal.
kuthi	a planter's residence
lakh	one hundred thousand
lathiyal	men armed with '*lathis*' or big, heavy sticks
latsahib	Lieutenant-Governor; a corruption of Lord + *sahib*
ma thakrun / kartri thakurani	mistress of a house; mother-in-law
mahal	administrative division of a landed estate; annexe of a large building
Maharani hawa	an atmosphere of admiration for Queen Victoria

mahashaya	a gentleman, an honorific term of address for a senior man, here the father-in-law
maund	a measure of weight (forty *sers*)
mauza	an administrative unit of villages
mondal	village headman
muhuri	a lawyer's clerk
mukhtar	lawyer who practices in a lower court
munshi	clerk
naib	an assistant, a deputy
namaskar	a greeting with folded hands
nazir	title of an Indian court official
nizamat adalat	court where criminal cases are tried
padri	missionary
pandit	a teacher of Sanskrit, a learned man
parwana	summons, warrant, grant
pathi / patinidar	lease of land with *zamindari* rights/holder of such a lease
peon	footman, orderly
pranam	a greeting given to an elder, whereby the feet of the person are touched
riayat / ryot	peasant
sadhu	an honest man
sadr	head, chief, the headquarters of a district
sarkar	the Government
sati	the custom of a widow being immolated along with the corpse of her husband
sepoy	soldier
ser	a measure of weight which varies locally
Shyama	an aspect of the Goddess Durga
Shyamchand	name of a whip used by the planters to flog peasants
siristadar	an office superintendant
taccavi (*taqawi*)	money given in advance especially to cultivators for digging wells, implements, etc.
taluk	heritable estate at a fixed rent payable, not to the landlord, but to the Government
tehsildar	officer in charge of administration in an unit known as *tehsil*

thakurpo	husband's younger brother (litt. 'father-in-law's younger son')
zamindar	landholder, landlord

Indigo and Colonial Bengal

I

INDIGO, COLONIAL BENGAL AND NEEL DARPAN: *AN INTRODUCTION*

In the early 1940s, a few years before India became independent, we often visited certain villages in Bengal to find some solution to their drinking water problems. Many of these villages were in the old indigo-growing areas of Lower Bengal, and as silent witnesses to a bygone era when white indigo planters ruled their empire within a much vaster empire with impunity, stood several dilapidated factories and outhouses. They stood rather precariously, on the edge of a village, on the bank of some river, on deserted fields. These 'little Englands', 'circles of improvement', as Governor-General Lord Bentinck (1828–34), bent on civilizing the natives liked to call them, were shunned by villagers as haunted places. Frail old men, in their late seventies or eighties, would tell us in whispers that ghosts of peasants who had died inside the lock-ups of the planters shrieked at night, and with a villager's faith in the divine dispensation of justice they would predict that as a punishment, rivers would rise one day and swallow up those remnants of tyranny and release the ghosts. A few stories, handed down the generations from father to son, almost like myths, could still be heard: of the planters' atrocities as well as of the bravery and sacrifice of Digambar Biswas who led the peasants in their struggle against the planters, and followed by Bishnu Charan Biswas and a group of other villagers, beat back the planters' musclemen, when they came to attack his village mounted on elephants. Despite these exhilarating tales the villagers had lost all interest in those oppressors; new, indigenous oppressors, no less ruthless had begun to take their place. The planters, if they lived at all, did so in the shadowy minds of a few old men; once they passed away so would the planters. All that remained then, to remind us of the physical presence of those white men, who had penetrated into the heartlands of rural Bengal, were those eerie ruins on windswept fields by solitary river banks. Even the indigo plant, which had brought fabulous wealth to some and untold misery to others, had disappeared from the soil of Bengal.

After our visit to those villages, marked by history, we decided to study Dinabandhu Mitra's *Neel Darpan* once again, to be able to perceive the reality of the oppressive world he had portrayed. We thought then that a modern, annotated translation of this play was required. But years passed, and it was only recently, after going round Bihar and West Bengal, that the extraordinary relevance in today's India of this century-old play struck us. We eventually produced this annotated English rendering of *Neel Darpan* (The Mirror Of Indigo).

THE PLAY, THE PLAYWRIGHT AND THE TRANSLATIONS

Somewhat as in the mid-nineteenth-century England, so also in the mid-nineteenth-century Bengal, 'the pressure of public demand for urban entertainment was causing brisk dramatic activity' (Sampson 1970: 613). In response to the mood of the audience, comedy, farce, satire, burlesque, extravaganzas and melodrama began to be produced in large measure. In early 1859 Michael Madhusudan Dutt's *Ekei Ki Bale Sabhyata* (Is This Civilization?), a biting satire on the ways and manners of the English-educated, westernized young Bengalis, was a sensation. The orthodox, conservative and wealthy Bengalis of Calcutta, who were the patrons of Bengali drama those days, hailed it as a masterpiece, and enthusiastically patronized its production. However, Dutt's second satire, *Buro Saliker Ghare Ron* (Old Sparrow Sprouting New Feathers on its Neck), a powerful dramatic caricature of the adulterous, pious-looking, orthodox aristocrats of Bengal, had no such luck. Finding a close resemblance to themselves in Bhakta Prasad, the principal character of the play, the patrons of Bengali drama were so furious that they prevented its being staged. From 1863 onwards Dinabandhu Mitra too tried his hand at comedy, satire and farce. *Sadhabar Ekadasi* (The Widow's Ritual Fast by a Woman whose Husband is Alive)—acclaimed as one of the best social satires of the nineteenth century and staged seven times in different parts of Calcutta within one year—was partly modelled on Dutt's *Ekei Ki Bale Sabhyata*. A shrewd observer, he must have realized that the days of tragedy were over: *Neel Darpan* was in fact his first and last tragedy.

Not long ago *Neel Darpan* was staged by a group of journalists in Bangladesh to inaugurate their union. The spectators were so entranced that no one noticed that in the suicide scene the man who acted the part of Gokulchandra Basu was really on the point of dying; he was taken down just in time and saved. Only recently the play was also produced—poorly, as feared—by *Doordarshan*, the Indian television. *Neel Darpan* has indeed been staged several times all over Bengal since the Great National Theatre of Calcutta produced it at their inauguration on 7 December 1872. In 1875, when it was staged in Lucknow, the rape scene in which the English planter was humbled by the Muslim peasant so enraged the European spectators that 'they crowded near the footlights. A few red-faced soldiers took out their swords and climbed on the stage' (Banerjee 1969: 186–7). The British magistrate cancelled the show and asked the troupe to leave the town. Amritalal Basu, a well-known actor and playwright, who acted the role of a woman, Sairindhri, narrated an amusing incident:

> Once when *Nil Darpan* was being staged, we heard that the Deputy Commissioner of police ... had arrived and many of us thought that he would arrest some (among the actors). None of us got dispirited, we got more merry instead. ... When the police boss heard this, he laughed and sent us the message: 'I know well Dinabandhu Babu. So I've come to see this excellent play of his. Why are you thinking otherwise?' (Mitra in Banerjee 1969: 186).

Evidently the reputation of *Neel Darpan* as a stage play spread early, and it was in demand. Though the system of buying tickets for witnessing a show had not been known before the National

Theatre now introduced it; there was no protest. The public paid willingly to see renowned actors such as Amritalal, Ardhendu Shekhar Mustafi, Kshetramohan Ganguli and others. The proceeds of the sale of tickets for the first two performances amounted to more than five hundred rupees (Guha-Thakurta 1930: 104). But the tremendous enthusiasm of the public proved to be a disaster for the play; it drew the attention of the colonial administration to its dangerous potentialities. The play's unobtrusive call to the common man to wake up to injustice and oppression appeared to them positively pernicious. Before long all copies of *Neel Darpan* were seized and its performance banned. The effects *Neel Darpan* had on the public each time it was acted speak for its immense contemporary suitability as a stage play, and do not corroborate the opinion of the eminent historian of Bengali literature, Prabhu Guha-Thakurta (1930: 109):

> *Neel Darpan* is an insignificant production. It is neither well-written nor does it lend itself to successful production on the stage.... Many of its unpleasant scenes, notably Act I (Sc. 3), Act II (Sc. 1), Act V (Sc. 2), are entirely unsuitable for dramatic representation.

All the famous stage actors of the time met every evening for rehearsals of the play; it is not likely that they would have staked their reputation on a piece which was not fit for the stage. The well-known writer Girish Chandra Ghosh, who in 1870 had acted the part of Nimchand in Mitra's *Sadhabar Ekadasi*, was to have appeared in the role of the planter, Wood, in *Neel Darpan*. In a moving dedication which he composed in the form of a letter in the memory of Mitra, Ghosh hailed him as the Master of Drama (*natya guru*).

While conceding its success as a stage-play, some are of the view that *Neel Darpan* cannot be rated high as a piece of creative writing. Its language is artificial, it abounds in ornate verse, it fails to create a world of fantasy, and instead of presenting truth 'through a veil of fiction', it records gruesome realities in their most naked and unvarnished form: '... horror piled on horror, and everything seemed to be found with slime and mud' (Guha Thakurta 1930: 109–10). Even if one accepts these criticisms partially, the question remains as to why *Neel Darpan* alone has survived, while its great contemporaries—Michael Madhusudan Dutt's *Sharmishtha*, or *Padmavati Natak* (Padmavati: Drama), Tarkalankar Ramnarayan's *Ratnavali Natak* (Ratnavali: Drama), or even Mitra's own *Nabin Tapasvini* (Young Hermit Woman) or *Kamale-Kamini Natak* (Kamale-Kamini: Drama)—are no longer heard of. These had all been rated very high for their imaginative qualities, their romance and flights of fancy. Yet, who today except the student of the history of Bengali literature remembers *Nabin Tapasvini's* romantic hero, Vijay, or his mother the pregnant queen, who left King Ramanimohan in anger, when he married again, only to return to him seventeen years later, dressed as a hermit? In this piece, fiction and fancy play their roles in full measure—the arrest of Vijay, his trial before the king, his restoration as heir to the throne, and his marriage to the beautiful Kamini, originally intended to be his old father's bride. The other great romantic hero Shikhandivahan of *Kamale-Kamini Natak* has also long been consecrated to oblivion. It was high romance, fiction at its very best, its setting far-off Burma, its hero the abducted son of the King of Manipur. The vicissitudes the hero passes through also have their horrors, but being entirely unrealistic did not shock public sensibility. Yet, neither Rana Kalyani, the beautiful, love-afflicted Burmese princess, nor the two delightful minor characters, Jaladhar and Jagadamba, full of frolic and fun, of *Nabin Tapasvini* can compete today with *Neel Darpan's* old, ugly maid servant, Aduri, or the poor day-labourer and simpleton in the planter's indigo lock-up; even the vicious Gopi seems to have a longer life than all the romantic heroes put together.

One characteristic of Mitra's plays is authenticity—which some have held up as a proof of his lack of imagination. In *Nabin Tapasvini* (1863), his first romantic comedy, some of the incidents, as in the story of Raja Ramanimohan and in the central plot of Vijaya and Kamini, were a mix of facts and folklore (Sen 1952–6: 59). His last romantic comedy, *Kamale-Kamini Natak* (1873), was based on the historical anecdote of Cachar (Sen 1960: 201). *Lilavati*, Mitra's third romantic comedy, is the story of the lost son of a rich man, who is reunited with his father after twelve long and hazardous years of separation. The main purpose of this play was to highlight the injustice of orthodox Brahmin *kulin* society towards women. Mitra may have taken some ideas from Ramnarayan's *Kulin Kula Sarvasva*, which had won the award announced by Kali Chandra Ray of Rangpur for the '... best play ... exposing the social evils of polygamy, and the iniquities of the Hindu social practice known as *kaulinya'* (Guha-Thakurta 1930: 56).

Lilavati, the daughter of the *kulin* Haravilas, is highly educated, and in love with a young man, Lalit. But she is forced to marry a debauch, Naderchand, only because he happens to belong to a *kulin* family. Since it is a comedy, everything ends happily to the delight of the spectators. Such was the demand for this play, that it was staged before packed houses for weeks every Saturday night, till bad weather intervened. In 1866 Mitra wrote a light farce, *Biya-Pagla Buro* (Love-Sick Man), also based on an actual incident in a small village. The story is about a doddering old man, who with a house full of grandchildren and a widowed daughter to boot, wants to marry again. Another, *Jamai Barik* (The Ideal Son-in-Law) ridiculed the custom prevalent among the aristocrats of Bengal of having live-in sons-in-law, like so many pets, so that their daughters did not have to leave their natal homes.

Whether it was his 'lack of imagination' or his lack of interest in Puranic stories and Sanskrit literature, Mitra did not produce a *Ratnavali Natak* like Ramnarayan, or a *Sharmishtha* like Dutt, or a *Sitar Banabas* (Sita's Banishment to the Forest) like Woomesh Chandra Mitra. With their basic difference in approach to dramatic writing—one culled his facts from contemporary society, while others preferred to turn to ancient literature for theirs—their mutual influence was negligible. As mentioned earlier, Dutt's *Ekei Ki Bale Sabhyata* did influence Mitra, but Mitra's young hero Nimchand, in *Sadhabar Ekadasi*, is an extraordinary and original creation. Nimchand comes from a good home and is highly educated in English, but he has become a debauch. He, however, never forgets that he is a man of culture, and this pride helps him keep his head high in a society where uneducated men lead drab and uneventful existences, never questioning the monotony of their lives. For such dull preachers of morality, Nimchand has nothing but contempt. This shows up at its best with the smug prohibitionist Gokul Babu: 'Can I join "society"? Be another Gokul Babu?' he asks. To stupid, impertinent Atal, the son of a rich man, Nimchand retorts: 'Had you a minimum of education, your words would have angered me, but now were I to get angry and argue with you, it'd be like honouring a stupid ass!' But Nimchand does not spare himself either; he is often the target of his own cynical jokes: 'I'm your Penal Code, Sir—You'll find every crime recorded there!' he tells the deputy policeman who comes to arrest him. But his jokes, the funny words of a drunken man, can hardly hide the pain of his wasted life, and though meant for laughter, they bring tears. Mitra had long given up writing tragedy, but Nimchand remains an outstanding tragic hero, who underlines the dramatist's compassion and understanding of human nature. Justice Sarada Chandra Mitra who, as a young man, in 1870 had gone to see the fourth performance of the play reminiscenced many years later in *Banga Darshan*:

> With the advance of age, I have forgotten many things, and perhaps I will forget many more. ... But I shall perhaps never forget Nimchand's performance that evening, and from that evening my respect for Dina Bandhu increased ... (Amarendra Nath Datta in Guha-Thakurta 1930: 98).

Mitra's interest in exposing social evils never left him, nor did his perception of reality. One of *Neel Darpan's* achievements lies in its portrayal of individual characters—of real people, who are no stereotypes, each different from the other. The loneliness of the young wife, Saralata, is very different from that of the widow Aduri, which again is not the same as that of the prostitute Padi'. Gopi and the steward are both despicable, yet not alike. Torap is an important character in the play; at a time when it was normal in Hindu society to think of a Muslim as being inferior to a Hindu, Mitra's Torap rises above all others in his courage, incorruptibility, and his devotion to the landholder's family. Torap has no parallel in the Bengali drama of the nineteenth century. Another unusual figure is that of Nabin Madhab; he is an activist as we understand the term today. Not a romantic by nature, he loves his wife, adores his mother and has great understanding for his young sister-in-law, Saralata. He is the channel through whom Mitra projects his own progressive ideas about women, who in the nineteenth century appear otherwise as beautiful queens, lovelorn damsels, or objects of lust. *Neel Darpan* was the first Bengali play to point out the low status of the women of Bengal.

If today we find the language of *Neel Darpan* artificial and at times pedantic, its verses ornate, we must remember the period in which it was written. Across the seas, in the land of the colonial masters, in the nineteenth century, plays and operas

> ... were vamped up from any handy materials. ... Plays were liberally interspersed with songs and dances. ... The special dramatic form evolved to fit the mid-nineteenth-century audience was melodrama, a term borrowed from the French.... Melodrama divided human nature into the entirely good and the entirely bad. ... Hero, heroine and villain, comic and virtuous retainers, heavy father ... fading and ultimately dying mother ... (Sampson 1970: 613).

An ocean lay between London and Calcutta, but their waves were splashing on our shores. In Bengal too those were the days of sonorous, rhetorical language, long soliloques, melodramas and stark tragedies. Plays without songs, verses and prologues were unimaginable: because the play *Sharmishtha* and Mitra's epoch-making comedy *Sadhabar Ekadasi* originally had no songs, Girish Chandra Ghosh composed songs for them and added a prologue to the comedies before they could be presented on the stage. Several decades lie between that world and ours, and it is not surprising that the demands and expectations those days were different from ours. As for 'horrors piled on horrors' in *Neel Darpan*, we would do well to turn back many centuries and reach back to the great Hindu epics Mahabharata and Ramayana. Another charge brought against Mitra has been that of 'flagrantly disregarding the dramatic proprieties' and using words and phrases that are 'obscenely suggestive' (Guha Thakurta 1930: 113). But if this is so he was merely following the tradition laid down by ancient Indian literature: the sexy dialogues in *Ubhayaabhisarika*, the erotic scenes in *Shakuntalam*, the scene of attempted rape in Mahabharata must then also hurt the sensibilities. Even allowing that literary merit be judged by standards of morals, double standards are unacceptable. In every age and clime a dramatist must cater to the taste of his public, if he is to be successful. In Bengal in the past the travelling companies of the *yatras* had catered to the non-elite; later the theatre increasingly filled this role. Even the National Theatre which was very elitist showed a preference for farces and Mitra's plays were an immediate success. Following his example pantomimes, burlesque, and farces, some of which depicted contemporary scandals, poured in and these were much more in demand than serious plays.

It is beyond the scope of this publication to analyse in depth the literary merits of Dinabandhu Mitra's work. The concept of drama as a serious criticism of life was not known in Bengali literature in the nineteenth century, nor can it be claimed that *Neel Darpan* introduced such a concept. But if realism, compassion, subtle humour, deep understanding of human nature and a vivid description of contemporary rural life are marks of creative writing, then *Neel Darpan* has high literary merits. Above all, perhaps, his recognition of the demands of the common people ushered in a new era for the Bengali stage.

The story of this powerful Bengali drama—Mitra's first venture in dramatic writing—is woven round the oppression perpetrated by a white Indigo planter in the 1850s on an old landholder and his family, the suffering of the peasants in Lower Bengal, and the tragic death of a young village woman after her rape by a planter. The play was published anonymously in Dacca in September 1860. Taking the anonymity further he explained in Sanskrit, in a short Introduction to the play, that some traveller had written the *Neel Darpanam Natakam* in order to comfort hundreds of villagers, lying miserably ill after the poisonous snake, the indigo planter, had bitten them. There was also a preface to the play, a strongly worded piece criticizing the planters, but expressing the hope that they '... would take a look at their own faces as reflected ...' in the play. The editors of the two pro-planter newspapers, *The Englishman* and *The Bengal Hurkaru*, were also castigated and compared to the corrupt Judas, who would do anything for money. It was perhaps this preface more than the play itself which stung the European community in Calcutta to the quick and led to the indictment of the Reverend James Long, who was held responsible for the translation of the work. Walter Brett, the furious editor of *The Englishman*, sent a copy of the preface to the Secretary of the

Landholders and Commercial Association of British India, and the passage was declared to be libellous. The prosecution charged that the preface had slandered the editors, and that the text of the play, a piece of foul and malicious libel, had slandered the planters as a group. Accepting the charge, the Chief Justice of the Supreme Court, Sir Barnes Peacock, found James Long guilty and sentenced him on 24 July 1861 to a fine of rupees 1000 and one month in the common jail. The trial and prosecution of Reverend Long of the Church Missionary Society by a biased judge and jury was such an extraordinary event that it has to be dealt with in some detail for a better understanding of India's colonial power structure and the risk that was involved in exposing unpleasant facts.

The first anonymous translator of *Neel Darpan* was Michael Madhusudan Dutt—the foremost contemporary poet in Bengali. His translation—entitled *The Indigo Planting Mirror*—was verbatim and several passages of the original, found to be 'objectionable' or 'coarse' by Victorian standards, were 'expunged or softened', and as Long mentions 'any one who will take the trouble to examine the original will find that a good deal has been omitted' (*The Calcutta Christian Observer*, August 1861). This early translation, deemed a classic, is a little quaint for the modern reader. We decided against a verbatim translation, since this would be difficult to follow for the general reader, especially for the reader whose mother tongue is not Bengali. Our translation diverges from that of Michael Madhusudan Dutt on a few other points also. An example is the name of the planter, whom Dutt called Rose; here, as in the original Bengali, he appears as Planter Rogue. The Bengali edition we have used to translate from is that edited by Brajendra Nath Bandyopadhyay and Sajanikanta Das and published by the Bangiya Sahitya Parishad.

As early as 1847 a first publication concerning indigo in Bengal appeared. This was W.S. Seton Karr's *Indigo in Lower Bengal*, published in the January issue of *The Calcutta Review*. In 1903 Kumud Behari Bose published his interesting compilation *Indigo Planters and All About Them*. The same year Lalit-Chandra Mitra's *History of the Indigo Disturbances in Bengal with a Full Report of the Nil Darpan Case* appeared. All these three valuable publications are over a century old and are not easily available.

Pradhan and Sen Gupta's *Nil Darpan* containing the play and a full report of the trial of Reverend James Long ran into three editions.

A new publication, we felt, should contain in addition to the play an analysis of the social, economic and political conditions of indigo production in colonial Bengal. To give the reader a clearer picture of the power of the indigo planters and their official Association, this volume also includes some details on the contemporary reactions of the planters and the government to the first translation of the play, an account of the trial and prosecution of the Reverend James Long and of the printer of this translation, C.H. Manuel, as well as the Address to Reverend Long by several prominent Indian citizens of Calcutta before his trial. Also included are the statement Long was not allowed to make by the Chief Justice, and reports of the trial in *The Hindoo Patriot* and *The London Times*. Some important Minutes of J.P. Grant, then Lieutenant-Governor of Bengal, and some excerpts from the evidence of the peasants (*raiyats*) and missionaries before the Indigo Commission have also been given.

Translating from one language to another is never easy. It becomes more difficult still if the two languages are different in structure and rhythm, and perhaps most important, reflect different cultures—the one in the East, the other in the West. Certain idioms, expressions and songs, which are part of the cultural discourse of one tradition, may well be unintelligible in another and might even sound absurd if translated literally without extensive notes. In *Neel Darpan*, for instance, no reader who is not acquainted with north Indian customs can possibly understand Sairindhri's words to Kshetramani, saying she may 'always wear vermilion in [her] hair' (Act I, Sc. 4). But this is a common way of blessing a married Hindu woman in Bengal, wishing her a long married life never having to face the misery of widowhood. Since the translator cannot tamper with the text to make the meaning clear, he has to explain in a separate note that the line of vermilion in the parting of a woman's hair signifies that she is married, that her husband who had first applied this vermilion to her hair at the time of their wedding is still alive, for it will be wiped out only after his death. Similarly to a western reader, who values most the privacy of a unitary family,

Sairindhris' prayers as a young girl that she may enter a joint family and have 'a mother-in-law like Kaushalya, a father-in-law like Dasharath, and a brother-in-law like Lakshman' (Act V, Sc. 2) would appear unnatural and even ridiculous. In order to appreciate this passage, familiarity alone with the virtues of these characters from the Ramayana is not adequate. The reader must learn something about the bringing up of a girl in traditional Hindu society. Or again the anguish of Nabin Madhab for not being able to celebrate the festival of Durga Puja properly might look somewhat odd unless one is told that in Bengal Durga symbolizes not only a goddess to be worshipped, but also the beloved daughter. Her brief, annual homecoming after almost a year's separation is looked forward to with emotional fervour, the re-union is moving and an occasion for great celebration. It is thus a matter of deep sorrow when this cannot be performed properly. Since such culture-specific references and passages abound in *Neel Darpan*, the translator is indeed in a dilemma: should the text be presented in modern English and left at that? Or should in addition to the translation extensive notes be provided for a non-Bengali reader to be able to enter into the spirit of the play?

Added to these problems is the rather unexpected one of understanding the language of the entire play, the variety of its vocabulary. This is a problem inherent in the very style of its presentation, which changes constantly like a chameleon. It is this which makes *Neel Darpan* vibrant and realistic in the original but puts the translator in consternation. The educated and well-read characters of the play, for example, often break into archaic Sanskritic prose, rich with similies and metaphors, replete with complex word structures. It is no easy task to reduce these long, meandering, sonorous sentences into shorter intelligible sentences, while keeping their contents and flavour intact. Then there is the British planter, showing off his supposed mastery of the Bengali language. The colourful Bengali swear words he loves to indulge in lose their peculiar, vulgar flavour when put into English; to quote a small example, the English equivalents of the terms *shala* (litt. brother-in-law) and *baeta* (litt. son) make no meaning in such a context of abuse. Also, should the English version of the planter's pidgin Bengali take into account the mistakes and loose constructions in the original?

Should it be King's English, or Cockney, or some slang or dialect of Lancashire, Sheffield or some other place the man—who is most certainly not upper-class—may have hailed from? But our most trying experience has undoubtedly been attempting to understand the cultivators and other village folk speaking among themselves in their own dialect. Dinabandhu Mitra, who was familiar with the dialect spoken in the rural areas of Jessore and Nadia in the mid-nineteenth century, used this with abandon. A large number of these words cannot be understood easily because of the manipulation of syllables. Initial syllables are dropped or replaced by others: the name *Ram* becomes *Am, ranga* (the colour red) becomes *nanga, rakta* (blood) becomes *akto, lajja* (shame) becomes *najja*. Difference in the pronounciation of consonants turns *nawa* (taking bath) to *naba* and *khawa* (to eat) to *khaba*. But both vowel and consonant change may take more drastic forms still, as in *katti* for *kaite* (to speak) and *kante* for *kotha theke* (whence). Although villagers from Nadia, Murshidabad or the 24 Parganas still use some of these words, many are no longer in use, and their meaning only emerges from the immediate context: the term *amabasya*, which means 'new moon', stands for the word *amashay* (dysentry) in the dialect of the districts; the same applies to the word *pill* which is a corruption of the English 'appeal'.

All translators face the difficulty of choice and unwittingly many tend to become creators of their own discourse rather than faithful renderers of the original. We too faced this problem. We wanted our translation to be readable, but above all we wanted our text to stick to the original without altering the ideas or dispensing with the local sayings, songs, idioms and mythological references. We have adhered to the text meticulously, but within permissible limits tried to clarify some points, so that they might be understood and appreciated by all. Certain words and idioms peculiar to Bengali, as well as the mythological references have been explained in separate notes at the end of each Act.

Neel Darpan has indeed been a difficult play to translate. According to our understanding, Mitra's aim was to portray the rural Bengal of the nineteenth century as he knew it—warts and all. As a result *Neel Darpan* mirrors that Bengal. As translators our aim has been to reproduce the image he had drawn—without altering or repainting it—the mirror of the mirror, as it were.

II

INDIGO AND COLONIAL BENGAL

Indigo is an almost forgotten plant; no cultivator in Bengal grows it today and no dyer misses its extraction any more. Although no one knows for certain when the cultivation of indigo first began in India, there seems little doubt that it is indigenous to this country and was referred to as such in ancient times. Pliny gave it 'the highest place after purple'; Dioscorides found 'indiens' a 'cure for ulcer' and also useful 'as a paint'. Marco Polo saw it being used by the people during his travels in India, and the Arabs knew it and gave it the Indian name '*neel*' (cf. Meyer 1908: IX). There is a detailed description of the preparation of the dye from the plant by Jean-Baptiste Tavernier (Vol. II): the sowing of the seeds, the cutting of the plants, their immersion in '*chuna* faced' vats, their transformation into paste and their finally being shaped into 'the size of a hen's egg'. Even the 'sifting' of the indigo to separate it from dust, and the care the people took to 'close all their orifices, only leaving two small holes in the cloth for the eyes' did not escape Tavernier. This locally produced indigo used to be sent to Europe by merchants long before the arrival of European planters in India. In Europe itself, woad, which was locally cultivated, was indigo's rival and held the fort for some time. During the reign of Queen Elizabeth I of England, for example, the imported indigo had to be destroyed (Choudhury 1965: 178n), and the same happened in Frankfurt in 1577, under the edict of the local government (Meyer 1908: IX). But after the establishment of commercial relations with Surat, in 1608, the East

India Company's Indian trade gave overwhelming primacy to indigo for at least a decade. In the first half of the seventeenth century an expanding market was created '... by the rapid displacement of the traditional dyes such as woad and cogwood ...' (Choudhury 1965: 176–7).

The Beginning

The history of indigo plantation in colonial Bengal is an example of how forces outside British control affected the lives and fortunes of hundreds of Britons and Indians. In the sixteenth century, India had been a major supplier of indigo to England *via* the Portuguese and *via* Aleppo (Krishna 1924: 94). *The Surat Letter* 31 December 1630 makes it clear that the English used to supply '... great quantities of indigo and calicoes to the Arabian and Persian markets...' (Krishna 1924: 95). All the principal markets in India were explored, and by 1625 Agra, Biana, Ahmedabad, Sarkhej, Jambusar, Cambay, Dholka, Lahore and the Coromandal coast had been tried. Interestingly Bengal's name does not appear among the principal markets. There was no control over the high prices demanded or the low quality of the dye, and serious complaints were repeatedly made of the mixing of dust: '... be sensible how that specie is one of the chief supporters of your trade ...' (Krishna 1924: 96). Added to this was the keen competition among the East India Company's servants, who were trading in their private capacity, inspite of the prohibition on doing so (Krishna 1924: 95).

India remained the main supplier of indigo till the last decades of the seventeenth century, when her inability to compete with the West Indies displaced her from the world market. Foster (1926–7 in Kling 1977: 17n) reports that as early as 1648 the Indian indigo trade was considered 'likely to decline ... because of the quantities made in Barbadoes, etc.' However, George Watt, writing in 1890, remarked that between 1664 and 1694 the East India Company carried out a lucrative trade in Indian indigo (cf Chanda 1984: 46). Be that as it may, in 1724 the East India Company abandoned the indigo trade (Sinha 1927: 178 in Kling 1977: 17n).

While the demand for indigo was rising because of the growth of the textile industry in England and its increasing use during the wars of revolutionary France, whose troops were clothed in blue, the supply of the dye from the traditional sources was drying up. The French indigo industry based on slave labour, mainly in its colony of St. Domingo, had been perhaps the biggest supplier of indigo to Europe; this was ruined after the slave revolt there in 1789. Finding the production of other cash crops, such as coffee and sugar, more profitable the British colonies in the West Indies had reduced the production of indigo by 50 per cent. After American independence the supply of superior quality indigo from Spanish held Guatemala and of middle grade indigo from southern Carolina ceased altogether (Kling 1977: 17). The lucrative trade was in peril, but Bengal came to the rescue and once again the fate of this plant was to change. The stabilization of British political control here combined with the appropriate climate and cheap labour made Bengal look highly covetable, and the Court of Directors of the East India Company turned to their abandoned indigo trade in India after nearly half a century. The cultivation of indigo in India by European methods began in the latter half of the eighteenth century. In Bihar (in Kanti, District Muzaffarpur) the first indigo factory was founded by a Dutchman, Alexander Niel, in the year 1778 (Mishra 1966: 133), but in Bengal most of the planters were British (Watt 1908: 668).

THE EARLY EUROPEAN PLANTERS

In the seventies of the eighteenth century, there was something of an indigo-rush in the inhospitable, mosquito-ridden villages of Lower Bengal. In *Neel Darpan* (Act V, Sc. 4) the frightened city girl, Saralata describes the pitch dark night, pierced by the eerie barking of wild dogs and the sinister howling of jackals from inside the dacoit-infested jungles of Swarpur. The planter 'James Hills, of Necchindipore, than whom a more highly honourable or more universally esteemed gentleman is not easily met with' (*The Times*, September 1861) himself issued a statement saying 'When I purchased the concern, the whole country around, and all to the north, was either a swamp or dense jungle, and but a few years previous to that it was a preserve for dacoits…'. When the early white planters began to enter the villages, the condition of those areas was indeed deplorable. Even in 1860 the Governor General felt that it was useless to employ special indigo commissioners or special district officers to go round inspecting troubled areas, because roads were very few, the land was a swamp for four or five months a year, and there were large unbridged rivers (GG 1860a). A correspondent of *The Hindoo Patriot* reported in September 1860 that '… moving about in parts of Jessore was impossible without the aid of a boat, lakes connecting marshes and rivers looked like an ocean' (Bagal 1953: Anonymous Letter No X).

Bengal was, in addition, still reeling under the ravages of the great famines of 1770 and 1788, '… whose ravages two generations failed to repair', as Hunter (1807, in Nag Chowdhury-Zilly 1982: 20) had observed. Cornwallis estimated that one-third of the country was in a state of wilderness, since the famine of 1770 had wiped out one-third of the population (Nag Chowdhury-Zilly 1982) and, '… in 1772, James Renell spoke of "whole tracts of land lying waste for want of hands to cultivate them" … a feature which apparently persisted for the next two decades …' (Datta 1989: 108n). There was thus, no dearth of wasteland. These planters must have seen wide belts of desolation, deserted villages, acres of waste lands and jungles, infested with wild animals. The 6,000 villages existing in Birbhum in 1765 had been reduced to 4,500 between 1771–2, and jungles appeared where villages had been. In Nadia how acute the distress was is revealed by the letter of the Resident about

> … the alarming prospect of the province becoming desolate from the flight of the *ryots* to the adjacent provinces, where they may be able to obtain subsistence by becoming day labourers … it will be very difficult to persuade them to return for the ensuing years' cultivation (Nag Chowdhury-Zilly 1982: 23).

Not finding employment, however, thousands of these uprooted swelled the already large numbers of Hindu and Muslim religious mendicants. But alms alone could not support so many thousands, and many of them began to raid villages; they demanded contributions from wealthy landlords and looted their homes if they refused to pay. With organized bands of their own followers and increasingly supported by starving peasants, disbanded soldiers and impoverished *zamindars*, these mendicants began to clash with British troops, and sometimes even defeated them. This mid-eighteenth century *Sanyasi Rebellion*, as it was known, was one of the most formidable the British had to face in the early years of their rule in Bengal (Fuchs 1965: 109–10). In fact, it could be described as one of the earliest uprisings of landless peasants.

After having been forcibly disbanded by the British, these men are not known to have gone back to their villages, as most of them preferred to work as mercenaries for warring princes. Thus the shortage of agricultural labour must have been considerable and was most probably one of the greatest problems faced by the early planters till about the third decade of the nineteenth century (cf. Datta 1989: 108n). This could well have been one of the reasons why the Frenchman Louis Bonnard, who settled in Chandernagore in 1777 and was reputed to have been the first planter to enter Bengal, failed to manufacture indigo in Taldanga village in the district of Hugli (Rainey 1873). The success of Carel Blume, another early planter, who had erected his factories in 1779, twenty-five miles from Calcutta, somewhere in Hugli and shipped his produce through the House of Paxton, might have been because of the huge sum (Rs 3,00,000) he is supposed to have spent. He sent a memorandum to the Governor General in Council on 12 July 1787 mentioning that he had tried as early as 1778 to cultivate indigo in order to improve agriculture and create a new article of commerce (cf. Sinha 1956: 206). The Board of Trade gave contracts to several contractors such as Douglas, Fergusson, Barretto, Henry Scott, and entered into its first contract with J.T. Prinsep in 1779–80, paying him an advance of Rs 1,70,000. He was to supply 2,000 *maunds* of indigo between May 1781 and May 1782 (Sinha 1956: 207). Prinsep tried to build a factory on the West Indian model, but failed (Kling 1977: 19n). Commercial residents of the East India Company were also permitted to manufacture and sell indigo to the Company or to private traders; one of these was Charles Grant, who planted indigo on his own in Malda district from 1780–7. The first Protestant Missionaries William Carey and Mr Thomas, who were looking after the factories of another commercial resident Mr Udney, in Malda '... viewed their engagement as a great opening for usefulness for the cause of Christianity' (Chanda 1984: 46).

Incidentally, for the greater part of the eighteenth century, not only civil servants and army officers, but '... even chaplains were engaged in trading' (Nayak 1981: 52).

> Till almost the middle of the eighteenth century most Bengal commodities retained a relative cheapness when compared to the same products from other countries. ... The British traders could balance the cheap transportation, abundant labour, very low cost of food grains and the more than enough land for expansion against the high mortality among European settlers in India, the hazardous navigability of the rivers and the diseases and discomforts (Nayak 1981: 52).

Of all other exportable items, such as raw cotton, sugar, hemp and flax, indigo had come to be considered the most suitable for the purpose of remittances to England, and it remained so even towards the close of the eighteenth century.

The Court of Directors expressed its hope that '... indigo will be a beneficient article of our commerce' (11 April 1785). A year later, on 12 April 1786, it observed (Letter No. 2): '... when we reflect upon the cheapness of labour in Bengal, and the favourable climate it enjoys, we cannot harbour a doubt of making indigo a most valuable article of importation ... we are confident that it might become one of the best means of remittances in this country'. But the company was in for a great disappointment regarding the performance of the contractors. They noted that, for example, the

> ... produce of the sales of the several parcels purchased of Mr. Prinsep prior to 1786 has yielded a remittance of only 37 pounds and 67 shillings, which reckoning the rupee but at 2 shillings is a loss in the first instance of upwards of 17 pc, independent of freight and charges, which may be reckoned at full 10 pc, more or less in the whole to a loss of 27 pc.

Throughout '... there were very heavy losses, independent of freight and charges ...' and it was suggested that '... the article must undergo a very considerable reduction in prices' (Letter of

28 March 1788 from CD to GG). The Court of Directors deplored that as a 'mode of remittance the indigo trade was found to be quite unpromising'; they lamented that the principal concern of their countrymen was 'to acquire property in order to return home' (Sinha 1956: 207). Since the system of limited contracting had not been successful, the trade was thrown open to experienced and professional indigo planters.

But where were such planters to be found? Among the requirements listed regarding the good management of indigo plantations were 'knowledge of the soil, of the language, dispositions, habits, connections of the lower order of natives' (Chowdhury 1964: 79). It was felt that 'few Europeans could ever attain such a familiarity of intercourse' with the natives as was required. Another difficulty mentioned was that 'arising from a want of capital, a want that in respect of Europeans continually recurrs'. Paradoxically, the Board felt that since

> the main object of the Europeans is to acquire property in order to return home, the moment ... the views of a successful projector be satisfied, he removes himself and his acquisitions and a needy man steps into his place.

Apparently, Europeans could not yet produce indigo as successfully and economically as did the local planters. The Board concluded, therefore, that the

> ... present desideratum for establishing indigo as a considerable staple in Bengal is that the natives should understand the manufacture of it so as to make it of a quality fit for the European market (B.T. Com 20 December 1791).

Further, the Board reaffirmed its conviction several times that if the natives could acquire the superior skill of the Europeans, there was

> ... no doubt ... that the simplicity and cheapness of their operations, the moderation of their profits which moves upon the scale that has produced the great improvement of Europe ... would enable them to sell it at such a rate as would afford a good profit to the merchant exporter and would probably gain possession of the market in Europe in preference to the Caroline or West Indian.

Why, then, were native planters not taken? No positive data being available, it may only be surmised that the native *zamindars* did not accept the Company's conditions, and chose to keep aloof. Though never a principal crop, some fifty years later indigo was manufactured on a large scale by several Bengal *zamindars*. Debendranath Tagore, for instance, had nearly 67,000 *bighas* of land under indigo; another big *zamindar*, Jaykrishna Mukherjee, had five to six hundred peasants working in his four indigo factories. References can be found to several other such planter-*zamindars*, such as Hur Kumar Tagore, Ram Jay Shaw, Fakik Shaw, Bhagawant Shaw, etc. (cf. Chowdhury 1964: 193). In 1854, of the indigo produced in India, two-fifths came from Nadia; some of it was also exported to other parts of Asia (cf. Kling 1977: 26n). But between 1770 and 1793 *zamindars* were passing through a crisis.

> *Zamindaris* were bought and sold, but what changed hands was not land *per se* but the revenue collecting rights and the financial privileges which accompanied such rights along with the trappings of prestige which attended the *zamindari* title (Datta 1989: 74).

Because of the excessive demand of revenue—it had been doubled—many of them were losing their lands at public auctions, and sent to prison as defaulters; agriculture was on the decline, and there was more land than men to till. With the arrival of the numerous indigo planters, the labour situation had worsened, and the competition among the rival planters to get hold of peasants and detain them till their indigo plants were harvested, often led to violent affrays and bloodshed. Whenever they could, the peasants either escaped taking advances from either of the rival planters, or cultivated only for one, though taking advances from both. In any case, the *zamindar's* own fields suffered; not seldom were their own lands also occupied forcibly and their peasants taken away. During those turbulent days the proposition of their growing indigo and selling it at a minimal rate to the Company could not have appeared attractive enough to the *zamindars* for them to withdraw their peasants from their rice lands or sugarcane fields and put them on to the cash crop indigo. This reluctance was most probably one of the reasons for indigo cultivation in Bengal very largely remaining in the hands of European planters.

Experienced planters from the West Indies were hence encouraged to come and were granted permission to establish their factories (Letter of 27 March 1787 from CD to GG.). One of them, Robert Heaven, who had 'for 13 years employed his talents in the culture of those articles in the West Indies ...' was permitted to proceed to Bengal and '... remain in the villages for 5 years' (Letter of 17 March 1787 from CD to GG in Council, Bengal). One of the 'talents' they brought with them was that of slave-driving. The Court of Directors of the East India Company expressed the hope that 'competition will not fail to operate in bringing the article to the greatest possible state of perfection as well as to ascertain the lowest rate at which it is possible to manufacture' (Letter of 28 March 1788 from CC to GG in Council, Bengal). This ruthlessly utilitarian policy was responsible for the evils in the system of indigo production in Bengal; it led to the merciless exploitation of the peasants and eventually to their uprisings.

The Company officials themselves feared that many of the planters were adventurers—Europeans of the 'lower orders' who had flooded rural Bengal. These were the *chandals* among the British, as Sabitri, the old wife of Golak Chandra Basu, calls them in *Neel Darpan* and these were the men who came in direct contact with the peasants. Many years later it was observed that the means they pursued to get rich quickly 'operated to the impoverishment and destruction of the country' (Letter L/C to GB, 27 March 1818), but so great was the need for '... a profitable external commerce as a medium of remittance from Bengal to England' (CEHI II: 315), that they were granted licences to reside in the interior of Bengal and given subsidies for their industry. Regulation 23 of 1795 permitted them to hold 50 *bighas* of land round their factory. The then Governor-General, Lord Cornwallis, granted them licences liberally, and the Board of Trade hoped that with

> ... proper encouragement given by grants of wastelands for the cultivation of the weed or means of the sale—their industry (would) soon kindle an industrial flame to the great emolument of those provinces and of the returns to England (Letter of 10 April 1811 in Sinha 1956: 206).

As long as the production of indigo was to the satisfaction of the Court of Directors and remittances could be sent regularly to England, the planters were certain that they would not be asked to explain the excesses they committed. These began as early as 1796; helpless *zamindars* complained to the Chief Revenue Magistrate (*sadar diwani adalat*) that planters had ploughed up half-grown food crops and sowed indigo there, had beaten up the peasants and carried off their bamboos, toddy trees and straw. Even when the peasants delivered the indigo they were not paid, and in desperation they fled the villages. Sir Thomas Munroe, one of the foremost administrators and one-time Governor of Madras, made some revealing statements about these adventurers in 1813:

> I find no difference in traders, whether their habits are quiet or not, when they quit this country (Britain). They are very seldom quiet when they find themselves among an unresisting people over whom they can exercise their authority, for every trader going into India is considered as some person connected with the Government. I have heard that within two or three years, I think in Bengal in 1810, private traders and indigo merchants have put the inhabitants of the country in the stocks, have assembled their followers and given battle to each other and many have been wounded (cf. Spear 1963: 165).

Apparently, Munro was alluding to the heinous crimes mentioned in a Government Circular, dated 13 July 1810, under the following heads (Spear 1963: 91):

> (i) Acts of violence, which, although they amount not in the legal sense of the word to murder, have occasioned the death of natives.
>
> (ii) The illegal detention of natives in confinement, especially in stocks, with a view to the recovery of balances alleged to be due from them or for other causes.
>
> (iii) Assembling, in a tumultary manner, the people attached to their respective factories, and others, engaging in violent affrays with other indigo planters.
>
> (iv) Illicit infliction of punishment, by means of a rattan or otherwise, on the cultivators or other natives.

But the authorities were indifferent; indigo was much more important to them than a few natives being beaten up or being forced to produce the crop without payment (cf. Sinha 1956: I, 209–10, Kling 1977: 39–40). For this criminality, the only punishment the Governor-General (1807–13), Lord Minto, could think of was to withdraw the licences of those planters against whom 'the offences had been established beyond all doubt or dispute'. Though there were 'numerous abuses and oppressions', the Governor-General was 'still willing to hope that this imputation does not attach to the character of the indigo planters generally'. He directed the Magistrate to destroy the stocks and to 'impress on all Europeans who wished to reside in the country the necessity of abstaining from ill treatment of the people'. The planters, however, were not 'impressed', and 'ill treatment'—an excellent euphemism for murder, plunder, confinement, assault, kidnapping and seizure of lands—continued for the next fifty years. Thus oppression became a habit of the planters, and crimes of 1810 were still being committed as late as 1860. John Peter Grant, who took over as Lieutenant-Governor of Bengal in 1859, mentioned in his Minutes (cf. Buckland 1901: 240) that

> the published records of Government show examples that have occurred within the last 18 months of each one of the four heads under which offences connected with indigo, as prevalent in 1810 are classified.

Basing his play on the evidence before the Indigo Commission and on his personal experience gathered during his official tours round the villages of the Districts of Nadia and Jessore, Dinabandhu Mitra showed how the two indigo planters, Wood and Rogue, were guilty of almost each of the crimes committed by the early planters around 1810. They were no figment of his imagination, and the stark reality of the presentation through visual symbols drew distant events painfully near, in a manner which press reports had failed to do. The horrified public of Calcutta now saw for the first time what happened to a village once it was leased to an indigo planter.

THE POSITION OF LOCAL *ZAMINDARS*

Some twenty years ago, a historian of rural India noted that

> ... there were indeed differences between what the nineteenth-century Englishman wanted to get and to do and what the nineteenth-century Indian landholder wanted to get and to do and ... there were also differences in the views of how land and labor could be used to achieve the desired getting or doing (Neale 1969: 8).

The early planters thus found to their dismay that a Bengal *zamindar*—often termed Raja—was not a landlord in the sense an English landlord was in their country; the former was like the lord of a territory or principality, whose right and duty it was to administer the country according to customary rights. The whole concept of territory, of land, was different, and even land measures were not measures of areas in the western sense,

> ... but a piece of land which satisfied the requirement that the *tenure holder* be able to farm some piece of land whose productivity accorded with his status rights; it was a 'fair share' as the people of that village or ruling brotherhood defined equity in relation to different families or castes (Neale 1969: 5).

Social stratification thus went hand in hand with economic stratification:

> Another feature of the *zamindari* was that it was highly stratified. At the apex were those who were, in the parlance of the East India Company, the 'principal' or 'capital' *zamindars*. These were the *zamindars* who had been able to carve out near-principalities and ruled over extensive territories. Positioned below them in the social hierarchy were the smaller *zamindars* with territorial jurisdictions over a few *parganas* (blocks of revenue paying units) and even over a few villages. ... This stratification notwithstanding, the *zamindars* were the most powerful non-peasant social block in rural society and also the most affluent. This affluence arose from their rights to tax and ... from their participation in agricultural production (cf. Phear 1872 in Ray 1979: 78).

Accordingly, the relationship of a *zamindar* with 'his' peasant was very different from that between the English landlord and his tenant; above all, the area under him was reckoned not in measures of land (*bighas*), but in the number of human communities (*mauzas*).

These early planters also observed the pomp and the princely style of living indulged in by the local *zamindars*, the appalling poverty of the peasants, and their oppression by their Indian masters. In fact this was no less severe than it was to be at the hands of the foreign planters; coercive methods, including flagellation and confinement in chains had always been used against the peasants for the realization of dues, and Babu Jai Chand Pal Chaudhury '... a great *zamindar*...' himself admitted that the peasants had always been compelled to plant indigo

> ... by numerous acts of oppression and violence, by locking them up in godowns, burning their houses, beating them, etc. (cf. Grant Minutes in Buckland 1901: I, 252).

This oppression continued and was also adopted by the European planters. It is interesting to note that it was on grounds of such oppression that the English exporters of indigo had, in 1633, refused to farm the indigo areas on the terms offered by one Manohar Das. The Moghul Emperor had granted the latter the exclusive right to buy all indigo in the kingdom in return for a payment after three years of Rs 6,00,000 out of his profit. The English merchants argued that

> ... whatever authority the king may give, it would be impossible for them to deal as rigorously with the people as their fellow-countrymen can (Foster in Misra 1961: 34).

The oppressive methods from which these English merchants shrank in the first half of the seventeenth century continued unabated throughout the nineteenth century and were now used by all.

The laws existing before 1793 exempted the owners of waste lands, cultivable wastes and deserted lands from paying revenue to the Company, and several *zamindars* were taking advantage of this law by literally forcing their peasants to desert and creating thereby acres of deserted land (Nag Chowdhury-Zilly 1982: 98–9). Fraud pertaining to land and revenue was universal:

Many ryotts in this district, thro' collusion with the *moffusul karmachari* [official in the village] have caused a considerable part of their *jummah* [assessed] lands to be set down in the accounts as *pulotoka* or forsaken [lands] & fresh leases at the lowest rate of assessment have been granted for lands to their own dependants, which is in fact to themselves. When they are called on for the payment of their original *jummah*, they produce these fictitious leases & thereby evade payment of a just demand (Datta 1989: 104n).

Such evasion was leading to a drastic fall in revenue collection, the organization of which was in complete chaos. The Company decided on various measures to curb these practices and obtain revenue, but these proved fruitless. Excessive increase in the amount of revenue demanded only led to '... distress sales of estates of defaulting *zamindars*..., very many of whom were eliminated in the process of the growth of a land market' (CEHI: II, 93), and as Sir John Shore noted 'I never know of any instance in Bengal in which the price of revenue lands sold, out of Calcutta, exceeded the revenue of two years' (quoted in Datta 1989: 104n). Philip Francis, a member of the Governor-General's Council' noted that by 1776 'the greatest part' of the *zamindars* 'were beggars or in debt' (CEHI: II, 93–4). The Report of the Collector of Midnapor corroborates this finding (Palme-Dutt 1979: 231). Eager to invest part of their money on landed property, buyers came from Calcutta and thus, '... the landed society ... ceased to be a closed one and any moneyed person could become a landed magnate' (CEHI: II, 93–4).

By the 1790s the Company had realized that arbitrary sale of rent-free holdings of *zamindars* and the stringent measures taken to curb their fraudulent practices were giving rise to more problems than they were solving. The revenue loss had not been made up and the expected revenue also was not forthcoming. Apparently to meet the revenue demand, or to make the best of the newly acquired property of the *zamindars*, the peasantry was being mercilessly extorted by the new buyers, some of whom were merely put up by British officials, who were the actual bidders. In a letter of 31 January 1776, the Court of Directors confirmed the fact of commercial infiltration of land, and noted that

> ... certain district officers held landed interest in the names of their *banyans*. ... Many of the most beneficial Farms in Bengal have fallen into the hands of the *banyans* and various remissions of rent recommended by the Gentlemen in the Company's service have been principally in favour of the Farms (Philip Francis, in Misra 1961: 124).

Few of these men had any experience in land management, and that the farming out of revenue collection by auction to the highest bidder was not a sound measure is clear from the discredit it brought to Hastings, who was Governor-General between 1774 and 1785. All this also often led to further peasant desertions, and as a result 'agriculture declined and the entire revenue system became unstable' (CEHI: II, 91). Added to this was the unholy emergence of a new group of large landholders, the *mondals* (village headmen), and employees of the *zamindars*. After the position of the *zamindars* weakened, these *mondals* increased their landholdings and often took over the lands deserted by the peasants at reduced rates. Though theoretically, a peasant could on return claim back his land, in reality this never happened. The *mondals* knew how to forge the land ownership documents (*pattahs*) in their own favour, and often controlled the land record offices (*patwari daftar*). Once the peasants' lands fell into their clutches, they remained there. 'The East India Company realized that in the name of peasant desertions and deserted holdings, a black market in land holdings ...' (Nag Chowdhury-Zilly 1982: 158–9) was being created. By 1793 the government felt that the old *zamindars* had to be restored, albeit in a greatly altered fashion; in spite of their impoverishment and weakened economic authority they were still the dominant social power in rural society. Thus, 'the social foundation of the new set up of 1793 was the old landed aristocracy only with a sprinkling of "new men".' (CEHI: II, 91).

The Impact of the Permanent Settlement

A new law was now called for which would enable the collection of maximum revenue with a minimum of effort on the part of the East India Company. Lord Cornwallis' Permanent Settlement of

1793 sought to achieve this objective. Under the new scheme *taluks* were separated from a *zamindar's* property; already under Hastings, the *talukdars* and smaller *zamindars* had received a distinct status (Misra 1961: 130). Now *talukdars* were recognized as independent revenue payers (Bhattacharya 1986: 115), and they and the *makarrardars* were recognized as separate from and independent of their overlords. The *zamindars* had hereditary proprietary rights, but the right of primogeniture was abolished in favour of the compulsory division of the property, in certain specific proportions, among all sons and the right to transfer land without the permission of the Court was given. It is then not surprising that estates started disintegrating. The *zamindar* was also deprived of his coercive powers over his peasants and tenure-holders. He was forbidden to demand the traditional tribute from them and was compelled to grant them ten-year renewable lease agreements (Bhattacharya 1986: 115). The total lack of purchasing power of the impoverished peasantry had begun to affect the market for British manufacturers, and this could well have prompted the unusual concern of the Company for the well-being of the peasants.

Above all, the *zamindars* were made responsible for collecting the required revenue, on the specific understanding that, regardless of drought, flood, famine or any other natural calamity, they would deposit in perpetuity a fixed revenue, the amount of which was either '10/11th of the current total payment of the cultivator' (Palme Dutt 1979: 232) or 9/10th of the net income, which in its turn was reckoned as 'the gross value of agricultural output minus circulating capital cost' (Bhaduri 1976: 46). The law made obligatory the payment of a portion of the revenue every month, and those failing to do so were imprisoned and their estates attached. There was a further understanding, that in case of default of payment their property would be sold; this sale would take place under the so-called Sunset Law, a cruel law demanding full payment of revenue before sunset on a particular date.

The new scheme consisted of a set of inflexible contracts between the government and each *zamindar*. Though several districts of Bengal had yet to recover from the devastating floods and drought of 1791, Lord Cornwallis made the excessive decennial assessment of 1790 permanent. As a result land sale began on a massive scale, and *zamindars* in acute penury could not support their peasants even with *taccavi* loans (Nag Chowdhury-Zilly 1982). In the first few years after 1793 defaulting *zamindars* were sometimes punished by imprisonment, in much the same way as they had been under the Moghuls (CEHI: II, 96). While the system of imprisonment was abolished in 1794 and monthly sales discontinued, a much heavier rate of interest of 12 per cent was charged on the arrears of revenue (CEHI: II, 100). The long spell of low agricultural prices between 1794–7 (CEHI: II, 94) rendered the revenue a crushing burden and it was then that the largest number of sales occurred. Unequal and excessive assessment of net income of estates—90 per cent of which were collected by the government—caused recurring sales by auction and this brought different groups of buyers, the indigo planter being among them.

The vital need of the planter was to acquire land and it must have been clear to those new arrivals in the villages, that once they could own land they could become planter-*zamindars*. Though the Regulation 38 of 1793 legally restricted 'Europeans of whatever nation or description' to 'purchase, rent or occupy, directly or indirectly any land out of the limits of the town of Calcutta without the sanction of the Governor-General in Council' (Sinha 1956: 27), and though Regulation 23 of 1795 further limited indigo planters to 50 *bighas* of land around their factories (Kling 1977: 55), there was no law preventing them from buying or taking leases of land in the names of their servants (Kling 1977: 55). As Colesworthy Grant (1860: 62–3) observed,

> The inability of European planters to hold lands in their own names ... occasioned a recourse to a system of fiction and concealment (land being held by Europeans under fictitious names—or in those of their servants).

Thus, through *benami* transactions planters could acquire several properties of small landowners, who often failed to meet the excessive revenue demands and lost their land. 'The Land Sales Registers revealed land sales of even small *Mauzas* belonging to petty landlords for minor rent worth Rs 122 or even Rs 100' (Nag Chowdhury-Zilly 1982: 10). Indigo production was hence not adversely affected and the planters could extend their indigo cultivation even before a later Governor-General, Lord Bentinck, opened the doors wide to them by legalizing the lease of land in 1830. James Hills, one such planter who came to Nadia in 1815, established his first factory in Nischintipur in partnership with the East India Company. By 1859 he had eleven such factories in the district of Nadia alone; he grew indigo on 67,000 *bighas* of land, of which only 14,000 *bighas* were not his own, i.e. were *beilaqa*. The Bengal Indigo Company, the largest indigo producer and landowner, grew indigo on a total of 75,000 *bighas* of land, out of which only 17,000 *bighas* were *beilaqa*. By 1860, as we shall see, the planters virtually owned two-thirds of Nadia District (Kling 1977: 54–5), which together with Jessore District, produced almost half the indigo of Lower Bengal (Kling 1977: 26). James Hills could thus boast:

> ... it has been kept studiously out of sight, how these two districts ... far outstripped and advanced ahead of all others in their prosperity. The answer is a simple one, the Englishman and his capital were there (*The Times*, 8 August 1861).

Already by 1854 two-fifths of all Indian indigo came from Nadia alone (Kling 1977: 26), and the village of Swarpur, which is the setting for *Neel Darpan*, may well have been in Nadia, from where Dinabandhu Mitra hailed, and to which he had been transferred as an inspector of the Postal Department, shortly before he wrote the play.

III

CHANGES IN VILLAGE SOCIETY

Meanwhile some significant changes were taking place within village society. The rural power structure had been undergoing a tremendous change and with the enactment of the Permanent Settlement, in 1793, the situation was to change even more drastically. Already by 1786, when Lord Cornwallis arrived, things had begun to change in a revolutionary fashion in the field of administration. Collectors and Magistrates had taken over from big *zamindars* their powers of administration. Now each of the major *zamindaris* of Bengal

> ... Burdwan, Birbhum, Nadia, Jessore, Rajshahi, and Dinajpur—were constituted as separate collectorates and smaller *zamindaris* and *taluks* were amalgamated into collectorates of similar size. Each *Raja* now found a British officer who was permanently stationed in the *sadr*, breathing down his neck, examining his accounts, ordering his officers, selecting his farmers, receiving petitions from his *raiyats*, in short telling him how to run his *zamindari* (Ray 1979: 80).

The authority of the *zamindars* had considerably weakened in the eyes of their own employees. Several peasants, most of whom were the *zamindars*' rent collectors, had enriched themselves; finding their masters deprived of their *taluks*, shorn of their official authority over the peasantry and compelled to deposit the revenue on an assigned date, at a particular time or else lose their estates, they refused to hand over the rents they had collected on behalf of these very *zamindars*. With these vast sums in hand, they themselves often became the buyers of these properties when they were auctioned under the Sunset Law. As Ratnalekha Ray (1979: 512) mentions, 'The largest amount of landed property belonging to the erstwhile Raja (Maharaja Krishnachandra of Nadia) was purchased by the local landholders of Nadia'. Large balance of revenue stood against every important *zamindar*.

> In Burdwan alone the number of suits for arrears of revenue from *ryots* ... was 30,000 in 1795. In Bishnupur it was the *mondals* who deliberately withheld payments (Ray 1979: 85).

This policy of the Government enhanced the power of the already powerful *mondals*. As userers and owners of acres of land, they ruled their peasants with iron hands and, as allies of the planters, they played a significant role in enforcing the indigo system. Planters realized how useful the *mondals* were and never antagonized them, used them as their contractors, whose business it was to distribute advances and supply the plants. It was much later that the planters fell out with the *mondals* and other employees, many of whom participated in the protest movements against indigo planters in the mid-nineteenth century. Chowdhury (1964: I, 170) has quoted the description of a *mondal* by Eden, the Joint Magistrate of Barasat in 1858:

> This man never sows a *cottah* of indigo himself; he makes all his *jotedars* sow five *cottahs* or a *beegha*, and thus produces the requisite quantity of plant, sends it all to the factory in his own name and takes the price himself and never pays a farthing to the *jotedar* who has actually sown the plant. ... In nearly every village in the indigo growing portion of the district, one *pucca* house may be seen. ...

The *mondal* in *Neel Darpan* was once an affluent person. He owned ten ploughs, forty to fifty bullocks and had at least sixty *jotedars* who used to be fed in his house, and the house itself had a courtyard huge like a racecourse, and a cowshed as high as a hill (Act I, Sc. 1). Later, when the *mondal's* two brothers refuse to sow indigo, they are beaten up; they leave the village and ultimately the family fortunes fall. Apart from the *mondals* there were the new or past owners of the defaulting *zamindars'* estates;

they were often employed by the planters in their factories as managers, overseers, clerks and other supervisory staff. These men also became useful supporters of the indigo system, even holding land on lease on behalf of an indigo factory. Most of them were Kayastha by caste, and of the twenty three 'leading families' in Calcutta in 1839 at least eleven were Kayastha (cf. Mukherjee 1970: Tab. 4). Indeed, the men who gained most in the New World (of British rule) were small traders, brokers and junior administrators, *pykars, dallals, gomasthas, munshis, banyas* and *dewans* (Mukherjee 1970: 43). These 'institutions', as Chowdhury calls them, 'were not actual producers of food crops, but they represented a scheme of distribution of peasants' surplus'. The planters, whose only interest was increased production of indigo, 'utilized them, made them a part and parcel of it and only thereby could [they] enforce a cultivation basically antagonistic to the peasants' production of foodcrops' (Chowdhury 1964: I, 170). Taken together, these groups constituted a kind of rural bourgeoisie. The formation of agricultural capital began with them, and the establishment of indigo factories served as a nucleus in that formation (cf. Misra 1961: 87). It was with the active support of these persons that the indigo system further intensified the already existing class-antagonism prevalent in the agrarian society. Mitra has described in *Neel Darpan* how the planter aggravates this hostility, often egged on by his high caste Kayastha employees:

GOPINATH: Enough of your high-flown eloquence, Sadhu! It sounds absurd coming from a mere peasant ...

WOOD: The sister-fucker's become a great pundit!

OVERSEER: He explains to the peasants the points of law and the meaning of warrants—this makes trouble for us....

GOPINATH: ... My lord, ever since the schools have been opened in the villages, the peasants have begun to give more and more trouble (cf. Act I, Sc. 3).

Planter Wood entirely agrees with his employees and promises to ask the Planters' Association to pressurize the Government to close down all such schools. A much greater man than Mitra's

Planter Wood—a prominent member of the Bengal Legislative Council—also felt uneasy about the

... general growth of intelligence regarding the rights of *Ryots* and other reflex results of education which has of late years pervaded many classes ... (Frere 1861).

RESTORATION OF *ZAMINDARS'* POWERS

But the '... awakening to the condition of serfdom', noted by Sir Bartle Frere was not in sight when, in 1799 and 1812, Regulations 7 and 5 respectively were enforced. With these, the short spell of respite from oppression was over for the peasant. Huge arrears of rent and the consequent fall in revenue collection apparently outweighed the Company's 'concern' for the peasant. Instead of reducing the excessive demand for revenue, which would have been the obvious step to take, the Government legalized coercive methods. The 7th Regulation of 1799 gave landlords the right to summarily attach and dispose off the property of their tenants for arrears of rent; they could even distrain the peasants' crop without going to Court. This 'replaced the old system of repayment of arrears through direct labour service by a more market-oriented system of attaching the tenant's property' (Sinha 1968: II, 160). It 'created a total and absolute property in land which had not existed before, and it accelerated the growth of a land market' (Mukherjee 1975: 43). The peasant had no redress against distraint and sale of his property, generally on false charges of revenue arrears, except to appeal to the Civil Court, which 'completed their ruin'. The Regulation 5 of 1812 rescinded the earlier privileges granted by the Permanent Settlement to the *raiyats*, of ten-year renewable leases and the liberty to come to an agreement with the *zamindar* (Ray 1979). Under this Regulation, *zamindars* had the full liberty to enter into contracts with their *raiyat* to their mutual advantage. But in all such contracts the peasant was always the loser. The Regulation 7 of 1799 was a boon to the *zamindar*. Exercising his new found power, he controlled those rich peasants and *mondals* who had cheated him of the revenue, and the collection of revenue improved. 'Compared to

1798–9, the outstanding arrears of revenue by 1804–5 decreased by 47 per cent' (CEHI: II, 317). Using their power of eviction the *zamindars* leased their land to new tenants on enhanced rent and a '... large number of under-tenures became a feature of the land system of Bengal ...' (Sinha 1968: II, 177).

A Hierarchy of Tenures

The *zamindar* had never had it so good when the subletting right of the superior tenure holder was legally recognized by the Patni Regulation 8 of 1819 (Bhaduri 1976: 47). The *patnidar*, having the largest number of tenure holders, 'enjoyed all the rights of the *zamindar* transferable and heritable with permanently fixed rent' (Kling 1977: 56). Like the *zamindar*, he too had the right to enhance rent and evict the under-tenure holders for non-payment of rent. The *patnidar* himself was bound by the same condition to the *zamindar* as the latter was to the Government: punctual payment of the fixed rent, and in case of default, summary sale of the land and termination of the lease. His interest naturally was to make as much profit as he possibly could, far exceeding the rent he had to pay the *zamindar*. The *patnidar* was a terror to the peasant, and in the following eyewitness account (cf. Nag Chowdhury-Zilly 1982: 121), a collector describes the relentless oppression perpetrated by one such *patnidar*:

> The greater part of the tenantry had been dispossessed of their lands ... the Patnidar had raised the rent of the rent-free lands of the ryots, and when they refused to comply with his demand of exhorbitant rent increase, the Patnidar supported by *Brijbasis* or dacoits forced the tenants to agree to the rent increase.... The extraordinary powers granted to the Patnidars to collect their rents gave them complete rights over their tenantry, who suffer from the forcible ejection of themselves and their families, and the over-assessment of their lands makes it impossible for them to take up their lands later if they return.

Another report says: 'The rapacity of the Patnidar has led to villages being depopulated, so that grass has grown in the houses, and many were destroyed by dacoits maintained by Patnidars'

(Nag Chowdhury-Zilly 1982: 121). Lieutenant-Governor Grant refers to this practice of taking under-tenures at excessive rents (Grant in Buckland 1901: 248, ¶29). Mitra describes the tyranny of the planters after they took *patni* lease of the village Swarpur. Their manager Gopinath Das assures the planter, Wood that, though a high caste Kayastha, he was in his work no different from a low caste *kaibarta*: 'I ploughed down Molla's paddy seedlings and had indigo sown there; I took forcible possession of Golak Babu's freehold orchard, which had been with the family for seven generations; their other rent-free property, which they've held since the rule of the rajas, is also in our possession' (Act I, Sc. 3).

The Regulation of 1822 further clarified the *zamindar's* power of eviction. If the peasant's tenancies originated after the Permanent Settlement of 1793, he was liable to eviction, and if he was not evicted he was liable to be assessed at the discretion of the *zamindar* (Mukherjee 1975: 93). Sometimes the *zamindar* allowed his own land to be sold for non-payment of revenue, and then with rigged-up sales bought up his own property at a very cheap rate, 'in order to wipe clean all leases and to enhance or evict at will as the new owner' (Baden-Powell 1892: I, 437).

The land rent under the Permanent Settlement fell heavily on the actual cultivator, who had many masters: sub-tenure holders, under-tenure holders, tenure-holders and also indigo planters, who held land under almost every type of tenure. The land '... let and sub-let half-a-dozen deep—and each holder in such case must have his profit out of this trafficking, the peasant of course finally pays for it all' (Grant 1860: 66). In some areas of Birbhum, *mondals* used the Law of Distraint to sell the peasant's grain, chattels and land, to collect the revenue arrears. Dispossessed of their few earthly belongings, and thrown out of their lands, they became landless. These lands, known as *palataka*, were acquired by the new landholders with the greatest ease. There is reference to such lands in *Neel Darpan* (Act I, Sc. 3): '... there's so much land lying waste that ... we can easily lease out another twent *bighas*', says Gopinath Das the planter's manager. The gulf between the peasant and his original *zamindar* grew wider and wider to become an ocean; the relationship that had grown over generations between him and his feared but accepted 'oppressor-*cum*-protector' ceased to exist. Assured of good income, some of these men lived in their estates, which were, however, managed by their agents or tenure-holders. The new relationship transgressed the old system of moral economy, and was now reduced to that between a ruthless employer and a resentful employee.

If the Permanent Settlement made the peasant's misery permanent, it also did cause some uneasiness in the Imperial mind. The land revenue being fixed permanently, the Government '... was no longer the primary gainder with a lion's share in the surplus appropriation, for the land revenue collected was appropriated by the elaborate hierarchy of intermediate tenure holders in land' (Bhaduri 1976: 48).

This might have been one of the reasons which prompted the Rent Act X of 1859, which 'legally fixed rent for those who held land at the same rate since 1793 or even for the last 20 years; right of occupancy was also granted to those who held land for the last twelve years' (Bhaduri 1976: 48n). But it was the Section XI of the Rent Act X of 1859, more than the Act itself, which made the planters bitterly resentful. This Section deprived them of their power to compel a peasant to be present at their factory, to force him to take advance, under the excuse of adjusting the rent due from him. Of all the powers which the Permanent Settlement had given to the planter, the power to compel the *raiyat's* presence at his *kuthi*, or office, was prized extremely highly—for this meant more indigo. The misery of the peasant in this situation finds mention in Mitra's play. Describing the force exerted upon the peasant, the defence lawyer in *Neel Darpan* says: '... when they go home grieving, there is such lamentation that one might think someone in the family has died' (Act IV, Sc. 1).

After the 'No-Rent' campaign had gathered momentum, Montressor, the Special Rent Commissioner, also mentioned in his report of 1861, that it was 'unquestionable that the adjustment of the rent account depended solely upon the *ryot's* consenting to grow indigo' (Montressor 1862). He further mentioned that

> refusing to receive rent, save at or near the factory is one among many difficulties in the payment of rent; the only conclusion I can come to is that the *ryot* had no means afforded him of settling the accounts, but by

the transfer of his balance from the *Zamindari* books to those of the factory as indigo advance.

After this Act was passed, James Furlong, the Manager of James Hills & Co., commented, that the result of implementing the Section 11 would be 'certain ruin to all the smaller class of talukdars throughout Bengal' (Choudhury 1964: I, 182–3). The measure was also criticized by Larmour, Manager of the Bengal Indigo Company, who believed that the *zamindar* had certain feudal rights, which would be destroyed by the Act. After Grant took over as Lieutenant-Governor of Bengal, important indigo manufacturers often sent to the Governor-General of India angry letters complaining against him and his officials. An extract from one such letter from Mssrs. Robert Watson & Co., Indigo Planters and Silk Manufacturers to the Secretary of the Government of India, is revealing:

This is a plain and literal account of our present position, and shows that practically our large business operations are completely stopped and the immediate ruining of our Indigo properties assured, unless it should please his Lordship in Council to intervene to remedy the mischief caused by the Government of Bengal. Mr Grant observed that all *ryots* are at liberty to claim such release from their engagements as Section 5 of Regulation 5 of 1830 can give them—the object of the preamble is amongst other things to afford persons, who may be unwilling to renew their contracts for the cultivation of Indigo. No proclamation can allay the ill-feeling that has been excited, or put a stop to the conspiracies or illegalities from which we are daily suffering. The combination is so universal and determined that from the convincing circumstantial evidence available, it is generally possible to afford the proof required by law, and no district officer can venture to accept less. ... The law does not suffice for the punishment of the evil doers (Watson 1859).

In 1860 Fergusson, who represented the planters on *The Indigo Commission*, gave a note of dissent, as he disagreed with some of the findings of the Commission; what he found particularly objectionable was that it tended to 'disturb the acknowledged principles of the Permanent Settlement, and to give *raiyats* notions of their rights incompatible with that contract between the

Government and the Zamindar' (Grant in Buckland 1901: 271). A few years later, the Secretary of the Landholders Commercial Association, in his letter to the Government, explained what Permanent Settlement meant to these landholders:

> The very essence and necessary condition of the right given to the Zamindar by the Permanent Settlement of 1793 was the landholder's power to deal as he likes with his own peasants, or in the case of land to let it only to those whom he chooses, and on such terms only as he pleases (cf. Chowdhury 1964: I, 181).

IV

INDIGO PRODUCTION AND FINANCE

Social and economic change was drastic in Lower Bengal between the parting years of the eighteenth century and the first decade of the nineteenth. Once sleepy villages were agog with *benamdars*, usurers, *mondals*, rent receiving intermediaries, tyrannical *patnidars*, rigged-up sales of huge estates for nominal sums and impecunious peasants, deep in debt and working like slaves in the indigo fields. There was a tremendous rise in indigo production from 4,952 *maunds* in 1788–9 to 62,500 *maunds* in 1795–6, when Bengal alone provided 67 per cent of the total indigo supplied to London (CEHI: II, 316). 'The cultivation and manufacture of indigo must be considered as an object of national importance' wrote the Court of Directors to the Government of Bengal in a letter (No. 10) dated 3 February 1796. By 1800, the Court of Directors found that '... the Indian Indigo had obtained a very general preference throughout Europe', and ten years later '... the article appeared to be established as a great staple of Bengal, and no rival to it seem(ed) likely to arise' (Letter of CD to GG Bengal, 20 June 1810). The long cherished dream of London becoming the 'emporium of Europe for the commodity' had almost become a reality. Delighted with such splendid achievements, the authorities did not feel inclined to probe into the actual process of this achievement. How the planters, struggling against shortage of labour and facing various regulations, could extort such abundant harvests was no concern of the administrators of India.

For the cultivation of the plant and manufacture of the dye the planters needed land, labour and capital. The three types of land which were generally found suitable for the cultivation of the plant were the high lands, as in Krishnagar, the low lands with alluvial deposit, as in some parts of Jessore, and the *chor* lands. The *chor*, or accretion lands were formed by a deposit of loam left on the banks of fallen rivers, after the waters had receded at the end of the rainy season, in October. Sowing began on these lands almost immediately before the mud grew hard, not by ploughing, but by merely scattering the seeds by hand—a method known as *chhitani*. October was the month for ploughing the low lands also. Cultivation on the high lands could not start before February or March, or even April, when the ground was warm and moist after the '... little rains known as *chhota barsat*' (Grant 1860: 94). The high lands had to be dug first, then ploughed several times and harrowed, and clods broken up and then ploughed again to be ready to receive the seeds.

The labour put in was hard and time-consuming (Mishra 1966: 336, Tavernier 1977: 8–9). In *Neel Darpan* Sadhucharan, a peasant, explains to the planter how impossible it will be for him to grow his own rice, if he is forced to cultivate indigo on an additional nine *bighas* of land: 'Indigo needs four times the care paddy does, so if I have to grow indigo on the nine *bighas* marked out, all my eleven *bighas* will lie fallow' (Act I, Sc. 3). The usually high and hard land had first to be dug deep, sometimes three or four times, using five or six ploughs at a time; even the smallest clod had to be broken by hand by the women and children before the sowing started. Once the plants came out weeds flourished, and again the entire family was out in the fields weeding, till the indigo was ready to be transported to the factory, either in carts or in boats—carts costing seven or eight rupees according to distance and boats four rupees per hundred bundles. After all this, generally, if at all the peasant saw money, 'the whole of the trifle of cash ... (was) given him in the form of an advance' (Grant in Buckland 1901: 263). Around 1840 an Englishman observed that

> ... after the advances are made, a man cannot sit down quietly in anticipation that the work will be duly performed, or the ground tilled by the stipulated time; either he or his agents must be constantly on the

alert to stimulate the people to work, and see that they do not squander the money they have received, or waste their time in going to some fair or festival, till the season is passed. In the old times, an Indigo planter would keep seven or eight horses in constant work in inspecting the lands, to the owners of which he had made advances ... (Frederick John Shore in Colesworthy Grant 1860: 77n).

Though there was an interval of five months between the two sowings, both the reapings took place at nearly the same time, the high land crops being ready to be gathered only four or five weeks later than the October crops, which were harvested after the winter. Both the reapings were done before the rains set in July, the plants reaching a height of five feet by the month of June.

There were two basic systems of indigo cultivation—the *nijabad* and the *raiyati*. In the former the planter grew his indigo on his own land and gathered his crops for his factory; he met the costs of production and faced the losses himself. In the *raiyati* system, the peasant took advances from the planter, and at his own cost cultivated the indigo on his own land; he harvested it himself and took it to the factory before fermentation began. If there was loss, due to inclement weather, for example, it was his bad luck.

After the plant was brought to the factory it was weighed and measured. The plant delivered had to be six feet both in length and girth; it was measured against a six-foot long iron chain, put around the girth of the plant. Whatever quantity the chain enclosed, after being pressed hard by the weigher (*ojondar*), became a 'bundle' (Day 1969: 209). Unless the bribe was to the satisfaction of the weigher, the plants measured into bundles of six or eight, before they were brought to the factory, were found to have shrunk to two or three. In the factory the plants were now put in the higher of two vats, which were always in pairs, with openings that could be plugged. These vats were usually twenty-one feet square and three and a half feet deep. The plant placed in the upper vat was then pressed by means of bamboos and large wooden beams placed on top of the vat, till the plant was at the same level as its rim. Through acqueducts and Chinese pumps, clean river water used to be pumped into the vat to steep the plant for twelve hours. When properly steeped the plugs were opened for the water to flow into the lower vat, leaving the dead plants in the top vat. These were then dried and used for the boiler, and as fertilizer. The liquid in the lower vat was then beaten continuously for two hours by the labourers—usually men from certain tribes—with bamboos, till the water, separated from the grains, was let out into the river and the soft blue grain was pumped into a huge boiler. The thick liquid was strained and pressed till it was solid and ready to be formed into blocks and cut into cakes. One such small cake, three and a half feet square and eight and a half ounces in weight, required one-third of an acre of land.

As for the capital required in the production of indigo, various legislations had been enacted in order to strengthen the hold of British capital. Trade flourished; the 'Kidderpore docks, for example, which had built 35 vessels in 20 years, between 1780 and 1800, launched another 75 in 5 years, between 1800 and 1805' (Moorhouse 1984: 74). The Charter Act of 1813 legalized free trade and allowed the entry of private adventurers. Even if the Court of Directors refused them permission, the Board of Control, on hearing their appeal, could over-ride the decision of the Court. Private Managing Agencies—a particularly significant factor in commercial growth—increased from fifteen in 1790 to thirty eight in 1813, and at the time 'there were 14 Scots, 10 English, 12 Armenian and 2 Portuguese merchants in the city' of Calcutta (Moorhouse 1984: 74). The financial assistance the planters received from the East India Company in the early days, or the loans they took from the Company's Commercial Residents, who also manufactured indigo on their own, hardly met their demands. The equipment needed to produce indigo was costly. They had to invest a fortune on pumps, boilers and other machinery and on the numerous out-houses—each one an indigo-producing unit. Here it was the Agency Houses—some fifteen Calcutta firms (Kling 1977: 20)—which came to their assistance and became their main suppliers of credit. In Bengal these houses controlled banking, investment, insurance and shipping and hence controlled a very great deal of private trade; they functioned as agents of the planters, who were settled in the provinces, handled for them the 'marketing and shipping of their products to Europe' (CEHI: II, 845), and invested in indigo a very large portion of their borrowed capital, namely the deposit of their clients' savings. 'In 1802, the private export of indigo, capitalized by the Agency Houses, was already more than three times as great as that financed by the Company's advances' (Kling 1977: 20).

With such ample supply of capital, the planters, though largely lacking in technical skill, set up numerous factories all over Lower Bengal. As a result, the production of indigo was colossal—40,000 *maunds* of indigo exported to England from Calcutta in 1800 rose to 1,20000 *maunds* in 1813, and it averaged 1,28110 *maunds* in 1826–30 (Chowdhury 1964: 83). By 1829 out of a total production of 1,49285 *maunds*, 1,08603 *maunds* were consigned to six big firms, which supplied one hundred and sixty lakh rupees out of the entire investment of two crore rupees. As the most secure medium of remittance, indigo became the principal article for investment of capital.

But these huge and seemingly impregnable edifices of British capital—trustees of a number of Indian businessmen—and considered to be '... as stable as the East India Company itself' (Mukherjee 1975: 41) had started developing cracks. In 1830–4 some of the leading houses went bankrupt. In 1827, a mournful letter from the Governor of Bengal to the Court of Directors of the East India Company mentions the cause of this impending disaster: 'No sooner than the capital accumulates from successful trade or agency than it is carried home by the possessor' (Chowdhury 1964: 85). This was further elaborated by *The London Times*:

> From each of these large Indian establishments, the partner has every two or three years advertised out and gone home to England with overflown fortune, being succeeded by another, who in his turn advertised out and went home very rich, and this arrived at by no other causes than the profits of their now bankrupt Houses, whose affairs, if justly examined, have been for years in a state of irretrievable ruin (19 April 1827, in Chowdhury 1964: 85).

The withdrawal of savings and deposits by the partners and clients of the Houses no doubt had caused an enormous depletion of capital, hastening their fall. The Agents, with their princely life-style and general extravagances—they were known sarcastically as 'the princes of the desk and ledger, with cotton as their crown and an indigo chest as scepter'—and the Directors too eager for inordinate gains (Singh 1966) were no less responsible for the catastrophe. 1826 also was a bad year for indigo; drought followed by heavy rains ruined the crop, and production fell by 50 per cent. The economic depression in England caused a further slump in the market. London prices dropped from 10 shillings per pound to 3 shillings 7 pence per pound in 1829 (Singh 1966), and the 'total price of indigo declined by more than a *crore* of *rupees*' (Chowdhury 1964: 85). Added to this were over-trading, improvident enterprise, extravagant miscalculation, cupidity and speculation.

Linked as it was to the Agency Houses, the indigo industry faced a short but real crisis; but being a prized commodity for remittances back home, planters did not have to wait long before a new group of Calcutta based firms, together with certain mercantile establishments in England, solved the financial difficulties of the indigo concerns. New factory blocks came up, there was no shortage of credit, and there was a sharp rise in the export of indigo. A letter (Fergusson 1838) from one I.H. Remy to the planter Fergusson mentions that

> the exports for the first three weeks have been 72,4911 lbs and for the month the total would be the largest we have had for many years, if not greater than ever before.... It will be at least 42,00 chests, about 1,000 chests more than in August 1837.

In 1839, the planters got an even more dependable source of capital—The Union Bank of Calcutta—which invested its own increased capital in indigo factories. 'The Bank would loan a planter up to two-thirds of the value of indigo in his godowns and the full value of his factory, house lands and buildings' (Kling 1977: 21).

The planters could never have expected a more understanding creditor; 'such loans', explained the Secretary of the bank, 'were to be drawn from time to time as [the planter's] necessities require, and repaid as his returns become available' (Chowdhury 1964: 113). In 1842 the statistics showed by the indigo factories had received more than half of the 'loans and credits on deposit of goods and other securities' (cf. Chowdhury 1964: 112). With unlimited credit from such a dependable source and with hardly any competition from indigenous traders, there was an unprecedented expansion in indigo cultivation. The result was overproduction and as a correspondent of the *Bengal Hurkaru* wrote, '39,000 chests, the fruits of surplus of 20 years encumbering the marts of Europe, stand like evil sentinels to paralyse the value of each new importation' (Chowdhury 1964: 116) The inevitable fall in prices depressed the indigo trade, and various suggestions—including that of burning down the the surplus stock—were made to combat overproduction (see Chowdhury 1964: 116 for some other suggestions made). A letter from a planter published in the *Bengal Hurkaru* (20 December 1847) also describes the desperate situation: '... unless means were taken to recover the trade from the present depressed state, the ruin to the planter ... which has been going on since 1841 must continue.'

In England also the trade depression was much more acute than in the 1830s; '... men who have lived twice as many years as I have, say', reported a journalist, 'that they never remember such a total disruption of the whole fabric of commerce—an earthquake as it were, engulfing so many of all classes—from the highest to the lowest' (*Bengal Hurkaru*, 25 October 1847). Prices in the London market, ranging between two hundred and twenty to two hundred and sixty rupees during 1822–6, had declined to one hundred and ninety-six rupees in 1841–2 and came down further to one hundred and sixty rupees in 1846–7 (CEHI: II, 318). Under these circumstances the majority of the planters were unable to pay off their debts, and major indigo brokers fell. Alarmed and anxious to terminate its indigo investments, the Union Bank now found that at least 90 per cent of its capital was tied up with indigo concerns. In December 1847 the bank failed. The comment of *The Englishman* (28 December 1847) was withering: 'It was acknowledged that all capital available for banking had gone and that the institution must be in future regarded as an indigo agency rather than a bank'. It might be interesting to quote *The Hindoo Intelligencer* (1847) describing the situation in Calcutta:

The Union Bank does not exist any more. There is no ship in the port. Only Chhatu and Latu (sons of the shipowner and a pioneer Bengali merchant in overseas commerce, Ramdulal De) are crying and rolling in dust. The Insolvency Court of Mr. Peel is thronged with people. ... There are brisk and false transfers of property on a large scale.

Local Seignories

Numerous factories, which had sprung up like mushrooms on borrowed capital, collapsed after the fall of the Union Bank. Big concerns, which dealt not only in indigo but also in silk, cotton and other commercial commodities, bought up these factories, going cheap and by 1860 they became owners of two-thirds of Nadia District (Kling 1977: 55). They acquired *zamindaris, taluks* and farm-leases, and after the formation of the Indigo Planters' Association in 1851, they 'honourably abstained from interfering with the portions of their neighbours, and partitioned out the country amongst them' (Buckland 1901: I, 241). Thus began the new regime of the planter-*zamindar* in Nadia and Jessore. Free to exercise 'permanently or temporarily a superiority in the lands around their factories', they had no rival manufacturers for miles around, no nagging worry of the cultivator being lured away by a neighbouring planter's promise of higher advance; in short, they became—to quote Grant (in Buckland 1901)—'local seignories'. But like all colonial masters, Grant too accepted the 'general view' that all this '... was as it should be' (Buckland 1901: I, 200). The Indigo Commission also agreed, and observed that 'there is nothing to complain of or amend so far as the tenure of land is concerned and the planter's relations with the native *zamindars*.' The Commission came to this conclusion after examining a number of *zamindars*, all of whom said in their evidence that

... they did not give *izarahs* of their estates willingly, but were compelled by repeated threats, and some by positive illtreatment at the hands of the Magistrates, who were on friendly terms with the planters, to give up their estates (*The Hindoo Patriot*, 25 June 1860).

Grant was not unaware of this, as he himself observed that '... native *zamindars* will never, till constrained by absolute ruin, sell their estates' (Buckland 1901: I, 147, ¶26). Commenting on the complacency of the Indigo Commission, *The Hindoo Patriot* wrote editorially on the

... extreme insecurity of all property situated within reach of the factories ... (where) the produce of orchards, and the preserves of fisheries are as much at the mercy of the factory people as the ploughed lands of the *raiyats* or their cattle and labour. But the fact has not sunk deep enough in the public mind. ... Few understand how frail is the tenure upon which they hold their landed rights, who are not prepared to wage eternal war with an enemy not more greedy than powerful, an enemy whom the law of the land scarcely reaches. ... But the men of property who stand in this danger have not made its existence so well known as it should be ... by their inertness they have allowed sad injury to be done to their cause. ... The evidence before the Indigo Commission teems with assertions ... that the *zamindars* ... misuse their powers ... they sell their *raiyats* and their planter-tenants from the basest motives, and they are too firmly possessed of their rights to be molested.

All this, however, was only to be expected. Lord Minto had foreseen this happening. Anxious to avoid the universal establishment of local monopolies of indigo manufacture, he had passed orders in 1811, rejecting the proposal that indigo factories should not be allowed to be established within 6 to 8 miles of each other. The following is an extract from this important document (cf. Buckland 1901: I, 239):

The natural tendency of such a restriction as that recommended by you would be to give a single individual an absolute monopoly of the produce of all the lands appropriated to the cultivation of indigo over a tract of country comprising many thousand *beeghas* round his factory, and consequently, to place the whole body of *raiyats* within that tract in a state of subjugation to that individual with respect to the price of the commodity. ... The *raiyat* would be precluded from that benefit from the cultivation of this valuable article ... which is essential for the amelioration of his own condition, and to the general agricultural prosperity of the country.

Lord Minto had something to say about the planter-*zamindar* relationship also:

It is not stated in your letter whether it is proposed that the *zamindars* and other holders of land, should be precluded from the privilege of establishing indigo works (as is done now by many of that class of people) within the stated distance on their own estates; if so, it would be a palpable infringement of their natural rights ...

Lord Minto's orders rejecting the proposal of the Magistrate of Jessore could not have been palatable to the planters. For years the established indigo planters had been opposing the setting up of factories by new comers, who had no scruples in giving advances to peasants who had already taken advances from them. They accused the peasants, who took money from both or more while working for one, of dishonesty. Several contiguous factories, competing with one another and each claiming the peasant for its own indigo fields, led to constant disputes and violent affrays. Even in 'February 1801, Mr Fletcher and twelve other proprietors of indigo concerns had appealed to Wellesley, the Governor-General, to prevent the newcomers by a "restrictive regulation" from interfering with their affairs' (Chowdhury 1964: 143). Another effort was made in 1804, but the Court of Directors 'enunciated the principle of non-interference in the relations between rival factories' (Chowdhury 1964: 144). The main concern of the Court of Directors being the total output of indigo, which was very substantial those days, and not the extent of profit the individual planter made, their reluctance to get themselves involved in the disputes of rival factories is not difficult to understand. In a letter to the Governor-General in Council, they did, however, express their concern over this matter (cf. Chanda 1984: 47). Since the proposal for local monopolies was rejected by Lord Minto, the Magistrate of Jessore now made two suggestions to bind down the fraudulent peasants (Chowdhury 1964: I, 148):

1. The planters be given the same power to realize their advances from the fraudulent *raiyats*, as the landlords had (under Regulation 7 of 1799) to distrain the peasants' crop for realization of their rent.
2. The planters be given the power the landlords enjoyed to summon the peasant for adjustment of their accounts.

This proposal too was not acceptable to Lord Minto. Advances, he urged, were like ordinary debts and recoverable on filing a suit in a Civil Court. The memory of the heinous crimes of 1810 probably influenced his decision in 1811. He declared:

Even at present, complaints are but too frequently proffered against indigo manufacturers for the violence of their conduct towards the natives ... the evil ... is at the present moment only partially eradicated.

Under the circumstances the Government '... could not discern the smallest necessity for converting the remedy already opened to the indigo manufacturer, of a suit in the Civil Court, into a criminal prosecution' (Buckland 1901: I, 242). Lord Minto observed further, that the powers asked for would 'open the door to the exercise of the greatest tyranny, oppression and exaction'.

LORD MINTO'S SUCCESSORS

The decade that followed Lord Minto's Government saw several changes, although his policy was continued by his immediate successor, Lord Muir, who was too preoccupied fighting the Gurkhas, the Marathas, the formidable Pindaris to take up the matter of the planters. Only in July 1823, towards the end of his term of office, and apparently influenced by the powerful Agency Houses, did he withdraw the Dacca Magistrate's order to the planters (11 June 1823) '... not to interfere with the lands of the *ryots* for cultivation of indigo, without obtaining previously the concurrence of *zamindars* and *cabouliats* (written agreements) from the *ryots*' (cf. Chowdhury 1964: 149). But the Agency Houses convinced the Governor-General that under such 'unauthorized and injudicious' restraints, no European indigo planter could possibly compete with native *zamindars*, who before long were sure to monopolize the indigo trade. Interestingly, on 7 July 1823, only four days after withdrawing this order and a few days before he left India, Lord Muir declared his opposition to monopolization of indigo:

Government cannot lend its countenance to a system having for its basis the assigning of the population of a certain tract of country to be exclusively dealt with by the owner of a particular factory. To adopt this principle would be to give a monopolising value to every existing factory ... success in the trade must depend on the influence created by the manufacturers' own character and conduct (Chowdhury 1964: 145).

Though the fall of the Agency Houses and the Union Bank was a rude shock to the business community, indigo planters had not been let down by the Government. On the contrary they had been taken special care of by the enactment of important legislations which furthered their interests. Nor had they been pulled up for their atrocities, which were being reported by responsible officers. To Lord Amherst the moralistic principles practised by his two predecessors—Lord Minto and Lord Muir—must have looked irrelevant, when he remembered the great commercial potential of the dye. Within a month of his arrival, in August 1823, without going into the root of the problem, and not even consulting his officers, he passed the Regulation 6 of 1823, which compelled the defaulting peasant *either* to pay back, in a summary suit, the advances he had taken from the planter and also meet the cost of the suit, *or* to bear in a regular suit all the expenses of the court case, and to pay back together with his supposed abettor—whose name might have been suggested by the planter—'equally and severally' the advances and the heavy penalties for not fulfilling the agreement. For no apparent reason it was assumed that while the peasants were so many frauds, the contracts they signed with the planters for sowing indigo were valid legal documents. If the contracts had been registered, as one of the magistrates had suggested, or the peasants been given legal rights to terminate their contracts after paying off the advances, instances of fraud would probably have been reduced. Years later, during the indigo uprisings, this very proposal of registration was again made by Lieutenant-Governor Grant:

The practices of charging the *raiyats* for a stamp without using any stamped paper at all, and of taking the marks and signatures of *raiyats* to blank papers, seem to me to require the check of registration. The registration should be so conducted that the identity of the person signing the engagement should be ascertained and capable of being proved at any time; and the fact that both parties fully understand all the conditions of the deed should be ascertained, and recorded by a disinterested officer (cf. Buckland 1901: I, 269).

But even in 1860, when the 'frightful length to which the factory *Amlas* were ready to go in perjury and forgery' was reported by two highly respected officers, Bell and Herschel, and the Lieutenant-Governor felt that all unregistered contracts for delivery of agricultural produce should be 'invalidated' (Buckland 1901: I, 269 ¶91), the majority of the Indigo Commission rejected the proposal.

In 1823 Lord Amherst felt that 'the most efficacious remedy which can be adopted is to provide a more expeditious and prompt mode of judicial process and thus to hold the cultivating *ryots* to the strict execution of their contracts' (Chowdhury 1964: I, 151). In 1824 he gave legal rights to coffee, cotton and sugarcane planters to lease land, but the Collector had the authority to withdraw their licences if they were found to have usurped peoples' property. The Board of Revenue also had the power to demand security deposits according to its discretion. It is possible that the serious nature of the numerous reports from his officers prompted Lord Amherst to impose these restrictions on the new planters. The Chief Judge of the Dacca Circuit Court of Criminal Justice, for example, had reported on 7 August 1823 that

some of the indigo planters and their native servants, with a view to enforce indigo advances to the *ryots* for the cultivation of indigo plant, frequently have recourse to the most violent measures to take forcible possession of their lands and to seize and confine the cultivators according to their options (Chowdhury 1964: I, 171).

Another judge from Dacca Circuit reported in 1824 that

sets of *lathiyals* were attached to almost every factory ... to enforce payment of outstanding balances from the *ryots*, to secure and hold their crops, and not infrequently to lay hold of and carry off the produce of neighbouring cultivators (Chowdhury 1964: I, 171; for an analysis of the *lathiyali* system in the *chor* land settlements see Zaman 1991).

LAWS AND THEIR ADMINISTRATION

The door, which had been opened partially to a section of the planters, was opened widely by Governor-General (1828–34) Lord Bentinck, in 1829; he granted indigo planters the legal right to lease lands and enjoy the right for sixty years. He also removed the few salutary checks imposed on them by his predecessor, Lord Amherst. The planters could now legally usurp others' property—something which they had been doing surreptitiously and illegally for years. In 1828, the Court of Directors of the East India Company had drawn Lord Bentinck's attention to reports of 'plunder, affrays, confining and assaulting cultivators and seizing of lands by planters against whom the local courts had proven ineffective' (cf. Kling 1977: 41). Despite these reports, Bentinck merely observed that 'this occasional misconduct of the planters was more than offset by the benefits they brought to the countryside' (cf. Kling 1977: 43). This opinion seems to have been fairly widespread among the British, and Colesworthy Grant remarked (1860: 81) that the negative behaviour of the planters '... is as nothing when contrasted with the sum of good they have diffused around them'. Bentinck took the view that the early legal restrictions prohibiting an indigo planter from possessing land had merely given rise to a '... system of fiction and concealment' (cf. Grant 1860: 63). While the Regulation 38 of 1793, which he must have had in mind, was indeed only an enactment on paper, hardly having been implemented, what impressed the Governor-General-in-Council was that, though the planters had 'laboured under severe legal handicaps, every factory [was] in its degree the centre of a circle of improvement' (Kling 1977: 43; cf. Grant 1860: 81).

After years of appealing and waiting for a special law, which would help prosecute a peasant criminally for the breach of a civil contract, the planters now got a more stringent law than they had ever asked for. Lord Bentinck passed the Regulation 5 of 1830, the harshest since the Regulation 7 of 1799. Under this new regulation, the breach of a civil contract sent a peasant to jail as a criminal, to serve his term there cultivating indigo. He was ordered to pay off all outstanding balances on the advance, but it was the planter's right to accept or refuse the sum. Further, the peasant had to prove to the satisfaction of the planter that the contract had ended; till such time the peasant had to work in the planter's indigo field. The Regulation was unjustly severe only on one party—the weaker one—for the contracts, held to be sacrosanct, 'were forced on an unwilling *ryot* by planter's *lathiyals* or forged by Indian factory servants' (Court of Directors 1828 in Kling 1977: 41). Apparently, it was this same practice which Lieutenant-Governor Grant was so keen to eliminate in 1860; he failed, but what did come to light then, much to the Government's discomfiture, was the involvement not of the factory servants alone, but of the planters themselves in forcing peasants to sign the contracts. When, for example, Herschel, a Government officer, was asked by the Indigo Commission to name the party who had informed him that the forgery, in which one Ganesh Ganguly was implicated, was committed with the knowledge of a certain Mr Furlong (an important planter and manager of twenty-three indigo factories), he said that he had been told by one Adam Smith that Furlong had insisted on a Mr Sanbelle proceeding with the false case (cf. *The Hindoo Patriot,* 25 July 1860). *The Hindoo Patriot* commented: 'Mr Adam Smith of course denied; who is to be believed? Mrs. Herschel? Or Mr Smith of Mohipore notoriety?'

After the passing of Regulation 5 in 1830, the peasant had theoretically only two ways open to him: either to continue to serve the planter till the latter agreed that the debt had been wiped off, or to file a suit in the Civil Court. In practice both courses were as good as closed. The first course was a blind alley for the peasant. The District Magistrate of Dacca did observe 'the utter impracticability of (the peasants) being able to free themselves, should they ever have the misfortune to receive indigo advances.' He wrote to the Governor-General:

The peasant will never earn enough to cover the meagre Rs 2/ which he received as advance per *bigha*; calculating the cost of cultivation ... per *bigha* as Rs 5 which included the Rs 2/ taken as advance, and even if production was 30 bundles—almost an impossibility—and the price of 8 bundles a *rupee*, the loss was *rupee* one and 4 *annas* (Letter dated 12 February 1830 in Chowdhury 1964: 131).

Another letter, written thirty years later, in 1860, by W.L. Hutchinson, an English settler in village Hanskhally of the largely indigo-growing Nadia District, testifies to the continuity of this system responsible for the appalling impoverishment of the peasant and his complete inability to pay off his debts:

Is the pretty appearance of lands near factories a criterion of the happiness of the *ryots*? Perhaps the planters may urge 'never mind fresh advances, let them pay us the old debts'. But the planters themselves have reduced the *ryots* to their present state of irretrievable indebtedness by advancing money on imprudent and usurious conditions. I can safely tell you that their (*ryots'*) chief demand is to be left alone, they do not wish to take advances nor enter into contracts for indigo leaf; so you need not fear of the *ryots'* 'repudiating their engagements'. It is the planter that is anxious to have the *ryots* do so, for it is less expensive (*Bengal Hurkaru*, 12 January 1860).

The second course suggested to the illiterate peasant 'to file a suit in a Civil Court' was like a mockery, reminding one of Lord Minto's ineffective 2nd Circular issued twenty years earlier on 22 July 1810. He had then directed his magistrates to '... report the proved instances ...', in which ... planters were obliging *ryots* to receive advances ... to compel them to cultivate indigo ...'. Neither the law of the land, nor the attitude of the magistrates had undergone much change during these years, and to expect the conviction of a guilty planter on the evidence of an humble peasant was simply absurd. The law itself was discriminatory; a Briton in India could be tried those days, be it for civil or criminal offences, only by the Supreme Court. Even after the Act XI of 1836 was passed, for criminal offences it was as before the Supreme Court alone which could try him, and though the Lower Court had the authority to try him for civil offences, the judge had to be an Englishman. With all these privileges, the British planter was *de facto* almost above the law, and beyond judicial control. For this minute loss of privilege, the Act XI of 1836 was denounced as a Black Act, by the English community at large, and interestingly, also by an elite Indian indigo planter—Dwarkanath Tagore—at a public meeting in Calcutta. Tagore accused the Government of trying to '... degrade the Europeans by lowering them to the state of the natives'. At a second mass meeting, citing examples of Europeans who showed great concern for the welfare of Indians, he justified his stand by pointing to the difference between 'interlopers' and 'officials' (cf. Kling 1977: 104–5). Apparently one important fact had escaped Tagore, namely that the law those days made not the slightest difference between the 'interlopers' and the Samaritans, as long as both were white. How intense the feeling of racial superiority was, is made clear by the following quotation, from a letter addressed to the President of the Board of the East India Company by its Court of Directors on 8 December 1831: 'The idea of being tried by Hindoo or Mussalman jurors must be intolerable to every Englishman.' In fact even after a decade of the passing of Act XI of 1836, protests from the English community—including the planters—were so violent, that a Government measure, intended to authorize Lower Courts to try Europeans charged with criminal offences, had to be withdrawn.

Even in the middle of the nineteenth century, although Lieutenant-Governor Grant felt that the increased number of Sub-Divisions had given the peasants access to magistrates, and that 'the protection of law was no longer hopeless' (Buckland 1901: I, 245), the reality was very different. In one case which came before the courts, McArthur, a planter of the Mirgunj concern on the Jessore-Faridpur border, had kept three peasants imprisoned in his factory lock-up for two months, because they had refused to cultivate indigo. However, McArthur was virtually exempt from criminal jurisdiction, while his servants were sent to prison. While hearing the appeal, Seton-Karr, the Sessions Judge, remarked:

> Looking to the difference of the Laws for Englishmen and Natives, and to the particular circumstances of the case, I cannot altogether reconcile myself to a decision by which the owner of the factory, an Englishman, should leave the Court with a fine, though a heavy fine (500 *rupees*), while a servant, a Native and an old man to boot, should leave the Court, not under fine, but for jail, under a peremptory order of imprisonment (Kling 1977: 82).

Another magistrate, Maclean, convicted a peasant sued by the Khalbolia Concern, who had been dead for thirteen years (Kling

1977: 159); when the error was discovered, Maclean simply dropped the case, instead of prosecuting the concern for perjury. Magistrate Moloney shared the Bench with planter Oman of Bijley, when he tried five villagers held for one month in a lock-up and accused by Oman of cattle-trespass. Magistrates used to stay with the planters when they came for inspection, and not surprisingly exonerated them of the most serious outrages, including murder (cf. Bagal 1953: Letter 28, 1860). One such case made a sensation, for it led to the transfer and subsequent resignation of A. Littledale, a 'judge of rare quality' (*The Hindoo Patriot*, 5 August 1861), for his dismissal of Planter White's appeal against a fine of three hundred rupees, imposed on him by the Magistrate of Nadia. The village of Govindpur, home of Digambar Biswas remembered to this day in Nadia for his courage to stand up against the planters, had been plundered by this planter's men and some villagers had been kidnapped. Tottenham, the Magistrate of Nadia, sentenced these men to six months rigorous imprisonment but fined White of Bansberia Factory, only three hundred rupees, although he was 'satisfied' that 'the whole circumstances of the case point to White'.

The system of one law for the Englishman and another for the native continued for years because of inherent racism in the colonial system, and especially because of the insistence of the powerful planters. As long as they had a white man to try them, they felt safe, and as Ram Gopal Ghosh observed, 'there [was] no practical remedy against the depredations and cruelties of European planters' (Majumdar 1963: 922). Even in 1883, the Ilbert Bill, which would have allowed Indian judges sitting alone to try Europeans, had to be modified following a storm of indignation. 'Gangs of indigo planters came in from the country', and 'along with their countrymen abused the Viceroy as his carriage passed by' (Moorehouse 1984: 81). Sir Alfred Lyall mentioned the reason for this indignation among the planters of Assam: '... they regarded the Bill as an attempt to do away with their right of beating their own niggers' (Majumdar 1963: 923).

If this was the situation in the late 1880s, it is not difficult to see why the weak had little chance against the strong in the 1830s. Those days there used to be only one Magistrate's court in each of the districts of Nadia and Jessore, from which many parts of these districts were sixty or seventy miles away. For a poor peasant to cover all that distance in order to file a suit against a powerful planter, in the court of a Magistrate whose language he did not understand and who did not understand his, and who invariably was a friend of the planter, was a nearly impossible, if not hopeless venture. The Court of Directors commented on the kind of justice an Indian could expect those days at the hands of the Magistrate:

> When the Indians complained, their petitions have been set aside for months; but when the Agency Houses (representing the planters) complained about certain suits going against planters, their petitions were acted upon immediately and the Government has gone so far as to interfere with the Courts in their cases (Kling 1977: 41–2).

Even in the 1860s the Lieutenant-Governor admitted that the 'chief fault was ... in the not impartial administration of the law' (Buckland 1901: I, 254), and though both Grant and the members of the Indigo Commission took the view that it was 'rather the system than the planters ... who are to blame' (Buckland 1901: I, 254), they could not exonerate the Magistrates of the blame that they did not hold 'the scales in equal balance', and 'have not accorded (the peasants) a due share of protection and support'. The Commission concluded that '... the practice followed by the executive authorities was favourable to the planters and hardly fair to the *ryots*', and commented on 'Mr Larmour, with two exceptions, up to this year' never having received anything but "support and counsel" from members of the Civil Service' (Majumdar 1963: 928).

After visiting the Krishnagar jail, where they found men convicted on false evidence and ruined through huge fines, and also two stone-blind men convicted of contract-violation (cf. Kling 1977: 149, 160), the Commission was convinced of the close co-operation between planters and Magistrates. They conceded that the Magistrate's '... bias has been unconsciously towards his countrymen, whom he has asked to his own table, or met in hunting fields, or whose house he has personally visited' (RIC Part I: 30). Mitra used these remarks regarding the Magistrate's weakness for the planters' table in a hilarious description of Sir

Frederick Halliday, the first Lieutenant-Governor of Bengal; one peasant asks another: 'how come the last Governor went about feasting with the planters, like a bridegroom before his wedding day?' (Act II, Sc. 1). Known for his extreme partiality towards planters, Halliday appointed twenty-nine of them as Honorary Magistrates. This had angered the peasants so much, that they had made a song which went *'je rokkhok, she bhokkhok'* (literally, 'the protector is the devourer'). Even the Commission noted that one Magistrate had sent a certain case to the 'very Honorary Magistrate' against whom the case had been filed (Majumdar 1963: 928). This extraordinary gift of the Lieutenant-Governor drew a scathing comment from *The Indian Field* (27 November 1858): 'This has infuriated the people of Bengal to an extent compared with which the greased cartridge of Hindoostan is a trifle.' The Governor-General of India also wrote to the Secretary of State disapprovingly:

> Let anyone who has read the evidence taken by the Commission say whether the position of an indigo planter conducting his factory according to that system and called upon at the same time to act as a guardian of law and as a protector of the weak from wrong, must not be a false, an untenable position (GG 1860a).

Halliday's proposal to re-enact Sections 2 and 3 of Regulation 5 of 1830 did not go through (Buckland 1901: I, 184). Those were tumultuous days, just after the Sepoy Mutiny of 1857; sporadic cases of resistance by peasants in various regions also had begun to worry the Government, and thus it was thought wiser not to provoke further crisis in the indigo districts of Bengal. Interestingly, however, planters in Bihar were given magisterial powers and posts in the army, in recognition of their fight against the mutineers, and their help to the Company in its restoration of rule (Mishra 1966: 335). Ritambari Devi (1983: 66) also mentions how indigo factories gave shelter to fleeing civil officers and their families; even the Joint Magistrate of a district fled Motihari, and moved to a nearby indigo factory.

SETTLEMENT OF EUROPEANS

In 1835 the Court of Directors, which never approved of the policies of Bentinck, repealed the Regulation 5 of 1830, by the Act XVI of 1835, and reduced the duration of the lease of lands from sixty to twenty-one years. Edmonstone, a prominent member of the Court of Directors, resented that: 'The momentous question of admitting Europeans and allowing them to become landlords was settled by the Governor-General without prior consultation with the Court of Directors' (Kling 1977: 44). Though the Chartered Act of 1813 had legalized free trade, every individual had to apply to the Company for permission before proceeding to India, the Board of Control being the final authority to accept or refuse the appeal of the individual. The East India Company viewed with disfavour any policy of unrestricted colonization; in 1818 it stated with a certain amount of unconcealed impatience:

> Every person who comes into this country is impressed with the idea of making, in a short time, a very considerable independent fortune. The means to be pursued to this end operate to the impoverishment and destruction of the country (EIC 1818).

But the Company's real cause for worry was not so much this impoverishment and destruction, as the possibility of losing larger taxes. Because

> ... the few wants of the Natives, who are satisfied with the mere necessaries of life, will allow them to pay larger taxes to Government from the same quantities of land, in the same state of culture, than an European can afford to do; it is evident therefore ... it cannot be for the interests of the Company to allow Europeans to become landlords (EIC 1818).

Bentinck, however, was a great believer in sharing the culture of the rulers with the subjects in India, and was convinced that this was possible by removing all those 'obstacles, which were productive of fraud, inconvenience and litigation—injurious to successful trade' (Grant 1860: 63). He was of the view that by allowing Englishmen to hold lands in their own names, by throwing open

India to British enterprise, British capital and English education, production of industrial goods would increase, costs of their export would be reduced—and the indigo trade would flourish. Spiritual enlightenment of the natives had already begun. Bishop's College was producing 'Missionaries and catechists for evangelizing the heathens' (Misra 1961: 150). Bentinck's views were shared by Charles Metcalfe, 'one of our cleverest statesmen', as Colesworthy Grant (1860: 63) put it. Metcalfe fully agreed that 'for the prosperity of our Indian Empire and to facilitate the settlement of our countrymen in India, conducive to the stability of our rule,' all those restrictions must be abolished. They were also held to be impeding 'that progressive increase of revenue which was essential' (Grant 1860: 63). Raja Rammohun Roy and Dwarkanath Tagore, two elite Indians of that era, endorsed these opinions of the Empire-builders; they were convinced that India's salvation lay in the settlement of the British in India. At a public meeting in England Rammohun Roy declared:

> The greater our intercourse with the European gentlemen, the greater will be our improvement in literary, social and political affairs; a fact which can easily be proved by comparing the condition of those of our countrymen who have enjoyed this advantage with those who have unfortunately not had that opportunity (Roy 1831; cf. also Roy 1945: I, 83).

At another public meeting in the townhall of Calcutta, Roy (1945: 15 December 1829) had made some interesting observations regarding the 'natives residing in the neighbourhood of indigo plantations'; he had found them 'better clothed and better conditioned than those who lived at a distance from such stations'. His comments on indigo planters were that there might have been 'some partial injuries done by some of them, but on the whole they have performed more good to the generality of the natives of this country than any class of Europeans, whether in or out of the service'. This speech was dismissed by *The Indian Field*, with the following sarcastic comment: 'What did Rammohun Roy know about the matter after a hurried tour through several districts? His remarks are no more entitled to reflect than the letters to *The Times*' (IPR: 27). Dwarkanath Tagore, who had established *The*

Car Tagore & Company, proprietors of coal mines, silk, sugar and indigo factories, and had risen from a '... *siristadar* to the salt-agent-and-collector of 24 Parganas ...' (Misra 1961: 127) to finally become one of the directors of the Union Bank, had little reason not to appreciate the benefits of close British co-operation. He himself observed that:

> ... through cultivation of indigo there is now not a *bigha* on the estate untilled ... several of my relations and friends have in like manner improved their property and are receiving a large income from their estates. ... if such beneficial effects ... have accrued from the bestowing of European skill on one article of production alone, what further advantages may not be anticipated from the unrestricted application of British skill, capital and industry to the very many articles this country is capable of producing (Grant 1860: 81).

This statement of Tagore is surprising in view of the fact that at least as far as one article—cotton fabric—was concerned, imports into India rose spectacularly from 1 million yards in 1814 to 51 million yards in 1830; by 1870 it was to rise to 995 million yards, thus driving out vast numbers of traditional Indian producers (cf. Kennedy 1987). Like Rammohun Roy, Tagore also found:

> ... the *ryots* materially improved in their condition, and possessing many more comforts than the generality of his countrymen where indigo cultivation and manufacture is not carried out (PSEI 1831).

Three decades later, the planter of the Nischindipore Concern in Nadia, James Hills, echoed the sentiments of these two Indian gentlemen. He told the correspondent of *The Times* in an interview (23 June 1861): 'Where abject poverty existed there is now plenty. There is scarcely a village in my properties where a number of its inhabitants do not pay income-tax.' This near identity of views had been possible because of very conscious and committed efforts on the part of the colonizers: 'We must at present do our best to form a class who may be interpreters between us and the millions whom we govern' (Macaulay's Minutes 2 February 1835 in Misra 1961: 154). But in 1831 the British magistrate of Benaras, Hamilton, did not accept the findings of Roy and Tagore; '... on the contrar...', he observed, 'I am inclined to assert the *ryots* are far better off in the *zamindaris*, where-in the proprietor will not hear of indigo being grown and resist every application for *patta* from factories' (PSEI 1831).

The link Bentinck had seen between the holding of lands by the British and economic advantage accruing from it was noted thirty years later by the Indigo Commission. It spoke of the economic importance of the indigo-trade in Bengal and the great political advantage of having a large body of European gentlemen scattered over the country (Buckland 1901: I, 256). It was noted with satisfaction that inspite of the uprising of the peasants, the indigo-trade could be 'correctly valued at nearly £2,000,000 Sterling a year' (Buckland 1901: I, 256).

Gradually, the interior of Bengal opened to the British. The Charter Act of 1833 continued the East India Company's rule for another twenty years, and subject to a certain amount of control, permitted British citizens to settle in any territory which had been under the Company in 1800. Four years later, the Act IV of 1837 made it lawful for any 'natural-born subject of His Majesty' to reside in, acquire and hold in perpetuity or for any length of time any part of the territory of the East India Company. The Company issued orders to permit grants of lands to Europeans provided '... they gave an undertaking that they would reside there and cultivate it ...' themselves, and satisfied the authorities that '... they had the means and intention of bringing to cultivation the land applied for' (Misra 1961: 213–14).

Acquisition of Land

After years of waiting, when European planters were permitted lands without legal restrictions, they realized that mere acquisition of land would not help them, unless they had the proprietory rights of the *zamindar* and his almost unlimited powers over his peasants. Henceforth their main preoccupation was to purchase or lease lands with such rights, wherever available and at any price.

... Buying off a *zamindar* he was simply purchasing the power of turning his *ryots*, so far as indigo was concerned, into slaves, who were expected to work not only without any share in the profits of their work, but at a real loss (Judge of Rungpore in Chowdhury 1964: I, 66).

It was, however, not always easy to persuade a *zamindar* to sell off his own lands, or even permit his peasants to sell their holdings to a planter. The big planter-*zamindar* Jaykrishna Mukherjee for example, refused on principle to grant lands to Europeans (Mukherjee 1975: 235). He told the Indigo Commission that he had invariably refused the applications which he frequently received from them for different kinds of leases, but had assisted them by getting his own peasants, in his own villages, to raise indigo for them on a limited scale. Answering another question as to whether he would permit or oppose, were an European to try to buy his peasants' holdings, he replied that he would oppose a new entrant, but would allow those who already had factories in his estates; they would have to pay the usual fees to him as *zamindar*. Mahesh Chandra Chatterjee, considered by the magistrate to be '... the most dangerous agitator in Nadia', and who was reported to have organized the disciplined demonstration by thousands of men, women and children in August 1860, when the Lieutenant-Governor Grant was returning from Patna to Calcutta, had refused *patni* leases to James Tissendie and Archibald Hills of Katchikatta Concern in 1859 (Kling 1977: 88).

The planters could and did go to any length when they were refused leases by comparatively small *zamindars*. Planters Wise and Glass, for example, kidnapped G.C. Chakerbutty, one such *zamindar*, in 1847 (RIC, Eden ¶5). Then there was *zamindar* Bijay Chaudhury, who in his evidence before the Indigo Commission stated that because he had refused to give lease to a planter, one of them had attacked his *cutcherry* and plundered it; then a false criminal case had been brought against him, charging him with '... burning the houses of coolies and looting the offices'. But the magistrate, being the planter's friend and having even stayed with him during his investigation, forced Chaudhury to negotiate, 'to make it up with the planter, otherwise it would not be good with him' (cf. Majumdar 1963: 937). For his refusal to renew the lease of his *mahal*, Jaggobandha Dutt, another *zamindar* was attacked by planter Maseyk of Kudumsar Concern. In his petition Dutt claimed that

> ... Maseyk attacked him with four hundred armed men, pushed in a wall of his house with an elephant, dragged out the women and children, sacked their jewels and one *lakh* of rupees and kidnapped him and another man, holding them for ten days (Kling 1977: 80–1).

The tragic case of Sital Tarafdar came to light only after the Lieutenant-Governor enquired into the petition given to him by the peasants of Nadia against the planter White of Bansberia Concern. It was discovered that Tarafdar had been kidnapped by the factory servants and had died in detention in a factory lock-up. This case might well have supplied Dinabandhu Mitra with the story of the death of Sairindhri's father in a factory lock-up (Act II, Sc. 3), and her anxiety that her husband might also meet the same fate. Mitra refers to several real-life cases in his play, many of which were reported by the anonymous correspondents of *The Hindoo Patriot*—he himself being one of them. In a letter (No. VIII 15 August 1860) it was, for example, reported:

> Mr Skinner (Magistrate of Jessore) is gone to Magoorah (Sub Division of Jessore) with 70 men of the Police Battalion to force the *Ryots* to cut the Indigo plant for the factories (Bagal 1953: 30).

There were complaints of the most serious nature against Skinner; he was devoted to a planter named Oman, and hence '... dismissed cases of arson and murder clearly proved against the planter' (Bagal 1953: 38). In a case in Mullickpore, 'though one man was killed, two kidnapped and four deeply wounded, none of the defendants [were] punished', since Skinner treated the whole affair as '... a common affray...'. In another case, looking into the complaints of petitioners—'... each one of them had a petition in his hand and then began to cry "*Dohre Dharmabatar*" '—Skinner arrested forty-nine of them: 'Chardon the planter, who sat by his side, began to point out and Skinner began to arrest' (Letter X, 3 September 1860). Skinner certainly served Mitra as a kind of model for his magistrates.

Instances of violence against *zamindars* by planters, however, decreased. By the nineteenth century a large number of *zamindaris* had undergone partition and fragmentation, and as we have seen, planters had become owners of two-thirds of the Nadia district. Apart from the fifty *bighas* of land around their factories, which had been given to them in 1795, they owned the *chor* lands, the holdings of absconding peasants (*palataka* lands), and of those who had died leaving no heir. They purchased waste lands of the villages and also the holdings of poor peasants. When they could not buy, they took long leases of villages, of private lands of *zamindars* (*khaskhamar*), and of low-quality (*utbandi*) lands often cultivated by the peasants following customary rights. The *zamindars* charged them excessive rent, '... far in excess of the gross rental lawfully demandable from the *raiyats*' (Buckland 1901: I, 243). The planters were far too shrewd not to have realized this, but agreed to pay the excess with the full intention of extracting from the peasants the same value in the form of indigo. This method was no different, as Grant commented, '... from an illegal cess ... only being more indeterminate [it was] obviously more liable to greater abuse in practice than a direct moneycess' (Buckland 1901: I, 249).

That the indigo system, in its various aspects, sanctioned abuse was not unknown to the authorities, but it was allowed to continue, because '... it has been so long in force, that it would be difficult to give it up without considerable loss, and in the great deficiency in the supply of indigo ...' (The Economist in *The Hindoo Patriot,* 23 June 1860). Since the peasant was unwilling to cultivate indigo, it was taken for granted that 'indigo can not be carried on without *zamindari* influence, and without oppression' (RIC ¶2928). So oppression continued in its various forms—physical, mental, and economic. Some, like Planter White, even used force to gain possession of the full quantity of grain available in the market and became moneylenders and grain dealers:

The planters thus accumulate the greater portion of the grain of the district in their *golabarees* (storehouse of grain) and the ryots' chief supply is obtained from these stores. If the planter refuses to sell or supply, the result is obvious (Report of the Magistrate of Nadia, 21 April 1860 in Chaudhury 1964: I, 185).

With the final emergence of local monopolies of manufacture of indigo, the peasant lost his advantage of inter-rural migration, which had been a kind of escape from oppression. There are references to such migrations in *Neel Darpan*: hoping to escape the cruel clutches of destiny, Sadhu Charan, a tenant, had migrated with his family to Swarpur village, but the oppression under the Swarpur planter became unbearable and he decided to leave again (Act I, Sc. 2); Torap, another cultivator, crossed the river at night to Basanta Babu's *zamindari* (Act III, Sc. 3). Apparently, though a *zamindar*, Basanta Babu was not an indigo planter. 'When the *raiyat* has a *zamindar*, who is not an indigo manufacturer, he has some protection in indigo matters. When the same man is indigo manufacturer and *zamindar*, or *zamindar*'s representative, the *raiyat* has no such protection' (Grant in Buckland 1901: I, 241). Whenever peasants sat together they discussed means of escape (Act IV, Sc. 1); the planter was also warned of mass migration if he continued to beat up his cultivators and increase the amount of their advances (Act I, Sc. 3). The play also refers to the desertions of villages by cultivators (Act I, Sc. 3).

THE SYSTEM OF ADVANCES

Advances, as a form of loan, were no invention of the planters. Wealthy men and women in the villages had always lent money to the poorer; there is a reference to this in *Neel Darpan* (Act V, Sc. 1) too.

But there was a difference in the manner in which advances were given by the planters. The Indigo Commission remarked that '... once a *raiyat* takes advance, he is never afterwards a free man'. However, it found no reason why planters should abandon the *raiyati*-cultivation—the primary base of the system of advances—and resort exclusively to *nij*-cultivation. Had the planters done this, the question of advance would not have arisen, since in the *nijabad*-system, the planter himself did the cultivation with hired labour. In the *raiyati*-system the factory-establishment absorbed the chief, or a large portion of the advance, a paltry sum of two rupees a *bigha*, which was given to the peasant when he signed the contract. He could hardly retain '... a half or one-third, or less than one-third of it' (J.P. Grant 1860: x). The meagre payment dwindled to nothing after the 'infamous *dasturi*' had to be payed, and various deductions made for seeds, rent, the stamp and the advance he had taken. The balance was always against him, and it began to mount in the factory books—how much he never knew for certain, being illiterate. All that he was told was that he was heavily in debt to the factory, and the little he had earned—'... the whole trifle of the cash ...' (Grant in Buckland 1901: I, 263)—was given to him in the form of an advance. There were '... lawful or quasi-lawful means ...' of pressurizing the peasant to cultivate indigo, and '... many written engagements contain[ed] a clause that any balance shall be paid, not in cash but in indigo, at the low rate fixed' (Majumdar 1963; X, 135). Even if some of them had to borrow from the *mahajan*, sell their bullocks or other property, they would, if they could, pay off the balance at once to get out of the factory books (RIC: Rev. Schurr ¶837). But this never could be; it was not permitted by the planters, and never would be: 'The planters would not clear them on principle, in as much as it would [have been] tantamount to closing the factory' (RIC: Cockburn ¶492). There was no escape. If a peasant refused to take the advance

> ... *lathiyals* [were] sent, numberless like locusts, his cattle are impounded, and carried to the factory, his ploughs are taken away, his recently sown rice is ploughed up and indigo sown by force (Rev. Joseph Mullins' evidence in The Economist in *The Hindoo Patriot,* 23 June 1860).

In 1856 one-hundred-and-fifty armed men were sent to coerce villagers, who would not sow indigo; three were murdered, villages were gutted, but Europeans were not punished (RIC: Eden ¶5).

The *raiyati*-cultivation on high lands was one of the reasons for discontent, and it looked to the peasant as if this system would be perpetuated. The Great Bengal Indigo Company, which already had 61,000 *bighas* under *raiyati*-cultivation, was contemplating further decrease in their *nijabad*-lands (Buckland 1901: I, 246). Big concerns, owners of thousands of *bighas* of land had already increased their *raiyati*-cultivation manifold. The Nischindipore Concern of James Hills in Nadia had 26,000 *raiyati* and a mere 4,000 *nij bighas*. Another big concern, the Sinduri Factory in Jessore, had 35,486 *bighas* under *raiyati* and 4,762 under *nij*-cultivation (Chowdhury 1964: 126). According to the planters in twenty-three concerns, *nij* covered 85,413 *bighas*, and *raiyati* 250,154 *bighas*. One quarter of the whole cultivation in these concerns was *nij* (Buckland 1901: I, 257). For the planter, *nij*-cultivation would have meant high production costs, engaging a number of supervisors, hiring out thousands of ploughs (two per *bigha* of land) and a large number of bullocks from hostile villagers. He would have had to employ hundreds of cultivators, and above all run around, over long distances, in distant villages, as *nij*-lands did not comprise one compact land mass. Neither threat, nor oppression would have increased the quantity of indigo, nor reduced the expenses incurred. Labourers would not have taken any advance, nor would they have entered into any kind of contract; so, even if the Regulation 7 of 1819 had been passed, these people could not have been punished like the mechanics, labourers and servants for non-performance of contracts. The *raiyati*-system, on the other hand, was infinitely cheaper. The *zamindari*-rights could be used with impunity to compel the unwilling peasant to take a small advance, sign a contract which, with no legal validity whatsoever, could be flourished as the perpetual sword of Damocles over his head. Under the contract, the peasant was to bear the entire cost of production, be responsible for the loss of the plant, permit factory servants to measure his land before cultivation, and deliver the harvested plant to the factory—all at a rate fixed by the planter. The advance-system could be so manipulated that the question of payment might not arise: 'Of 33,200 indigo raiyats, who cultivated for the Bengal Indigo Company's concerns in 1858–9, only 2448 were shown by Mr Larmour to have received any payment' (Buckland 1901: I, 252). Apparently, as the Commissioner of the Burdwan Division had stated, '... the wealth of the ryots is inconsistent with the well being of the factory ...' (Chowdhury 1964: 164). The balances piled up for thirty to forty years, and according to Larmour, had reached the fantastic figure of seven lakhs and seventy-eight thousand rupees. As Chowdhury (1964: I, 156) noted, 'these never-to-be-liquidated balances represented an inexhaustible reservoir of labour-power ... which planters could tap at any time'. The *raiyati*-system had one other advantage for the planter; it concealed unpaid labour. It was not the *raiyat* alone who tended the crop; it was his entire family.

V

GROWTH OF DISCONTENT

There was discontent, and it was growing. As Sadhu Charan puts it succinctly in *Neel Darpan*, '... there are things one can do and things one can not. Ram an eight-finger-long charge into a half-finger-long gun barrel and it'll burst' (Act I, Sc. 3). In rural society it had been normal for a poor peasant to ask his village landlord, who was often also a money-lender and grain-dealer, to give him some advance. He needed this, either in cash or in kind for marriages and funerals, for various seasonal festivals, which were part of his life, and during droughts and floods when his crops failed. Once he got the advance or the loan, at a very high rate of interest, he was usually left alone; generally there was no interference with his crops. In course of time, if he could pay back the advance with the promised interest, the cash was accepted and the whole transaction was clear. However ruinous the interest, and however great the terror the money-lender invoked in the heart of the peasant, this was a socio-economic pattern of exploitation both had learned to follow since time immemorial, and it was part of the tradition to which they were accustomed. The planter-peasant relationship, however, could not be located within this tradition.

In the 1850s the planters insisted on the old rate of payment, although the prices of agricultural produce in general had almost doubled. The peasants were illiterate, but not ignorant, and they were realizing that it was their own labour which produced the excessive wealth of the planter and increased their own excessive poverty. When the price of agricultural commodities rose, had a peasant the freedom to grow his own crops the net profit he would have made after its sale at the market price would have been considerable; instead, as Grant observed, he was put to a loss of seven rupees per *bigha*, cultivating indigo (Buckland 1901: I, 249). Even Larmour, the Manager of the Bengal Indigo Company, admitted this:

If *nij*-system produced 24 bundles a *bigha* of high-land, and the cost was Rupees 6-14-6, and if a peasant had to incur such an expenditure, taking the rate of six bundles a Rupee, his loss would be Rupees 2-14-6 (RIC: 1860 Appendix ¶22).

Another planter admitted before the Indigo Commission that the peasant lost eleven to twelve *annas* per *bigha*. Grant himself came to the conclusion, after some calculation, that the average price fixed by the planter never went beyond two rupees for four to eight bundles of indigo per *bigha*. The peasant had to pay four to eight *annas* for seeds per *bigha*, which he had to buy from the planter; the stamp paper on which he was to sign his contract cost him another four to eight *annas*, though most of the time he signed on blank sheets and was cheated (Montressor 1862), the stamp money going in to the pocket of the planter's men. Then the cartage cost him four to seven or even thirteen *annas* per *bigha* of product. The average rice crop was selling those days at fourteen rupees in Jessore, and eight rupees and ten *annas* per *bigha* in Nadia; at the latter place its price in the husk had risen from ten *annas* per *maund* in 1850 to one rupee and six *annas* per *maund* in 1855. For a cultivator whose main sustenance was rice and who had a family to feed—'... four *ser* of rice to start the day ...' (Act I, Sc. 2)—buying it at the prevailing rate was quite beyond reach. The price rise had drastically affected every small item needed in a peasant home. For lack of straw, huts could not be thatched, cattle could not be fed; linseed also had risen to ten rupees per *maund*. Even the repair of ploughs cost much more than before. According to the set of cables published by the planter, James Hills in *The Times* (23 September 1861), the price of

... ordinary crops like mustard seeds, linseed, cotton, sugar had increased 100%. Rice, wheat and grain had more than doubled; turmeric had more than trebled, and even ginger was fetching a price 20% higher.

Fear of starvation had gripped the peasants, because they lived so close to the margin. '... we better hang up our pots and pans' (Act I, Sc. 1) are desperate words; there was nothing more to lose and less to gain, for life had become one continuous pattern of misery. These words of the peasant Sadhu Charan reveal the stark situation peasants were facing. Under the prevailing system 1/16 of their best land, marked out by the planters, had to go under indigo; and then began the ceaseless, thankless labour of cultivating a plant which brought them nothing but misery. Although Lieutenant-Governor Grant (Buckland 1901: I, 250) observed that 'These *raiyats* are not Carolina slaves, but the free yeomanry of the country', W.E. Latour, the Magistrate of Faridpur, another important centre of indigo, when appearing as a witness before the Indigo Commission called the entire system of indigo plantation '... a system of bloodshed' (Mitra 1906: 3). But the changes required to be made in the system, as between the manufacturer and the cultivator, were more or less left to the heads of the various indigo concerns themselves. The planter James Hills, for example, felt justified in having raised the rent of his peasants by

> ... only *rupee* one per *bigha*, though their income in my estate is no less than sixteen *rupees* a *bigha*, and in many instances more. ... Although according to custom the share of a landlord in the produce of the land is generally one-half, but not less than one-third (*The Times*, 23 September 1861).

The correspondent of the paper commented, 'He may well assert, this moderate assessment is an act of indulgence on his part, yet it is this that has excited his *ryots* against him, and induced them to combine to refuse all payment'. When the peasants sat together, discussing ways of escape (*Neel Darpan*, Act IV, Sc. 1), these must have been the thoughts which raised storms of anger and antagonism, and urged them on to action.

Peasant Resistance: The Beginnings

Throughout the nineteenth century, even during the harsh years when Regulation 5 of 1830 was in force, one finds instances of peasant resistance—sometimes in violent forms. In the middle of 1829 Pargannah Haveli, in the district of Dacca, was convulsed by just such an affray; the indigo cultivators stood up for their rights, and it was only with the help of special police and military force that the situation could be brought under control by the rulers (for details cf. Chowdhury 1964: 193). The Fara'izi movement (cf. Ahmad 1966, EI 1913, Fuchs 1965: 116–24) has also been interpreted by some as having contributed to organizing peasant resistance. This movement was founded by a puritan Sunni, Hajji Shari'at Allah in village Bahadurpur in Faridpur District with the intention of reforming Islam, strongly contaminated, he felt, by Hindu practices and rituals; it took a more openly political turn under his son Muhammad Muhsin, better known as Dudu Miyan. A firm believer in social equality, he fought against economic exploitation and oppression of the peasants, and frequently came in conflict with *zamindars*, indigo planters and the Government which supported them. He urged his followers coming from Dacca, Bakarganj, Faridpur, Pabna, Barasat and 24 Parganahs, to resist oppression, refuse to pay taxes, boycott British Courts and settle all their disputes in their own councils (*panchayats*). The *zamindars* and indigo planters had been after Dudu Miyan and had had him arrested several times since 1838, but failed to get him convicted for lack of evidence. In 1847 the Fara'izis set fire to one Mr Dunlop's Cossimpore factory, and carried off the planter's right hand man, Prasad Kanai Lal (RIC 1860: De Latour). At last, in 1857, they succeeded in putting him in Alipur jail; Dudu Miyan died in Bahadurpur, in 1860.

Also in the 1830s, the Wahabi movement under a man called Mir Nasir Ali—and known to this day in West Bengal as Titu Mir—had taken a violent, anti-*zamindar* and anti-indigo-planter stand. This movement, known also as Titu Mir's Barasat Rising (Ahmad 1966), was directed particularly towards those areas in Nadia and Barasat '... which were studded with indigo factories'. Titu Mir, who was born in the village of Chandpur, now in

Bangladesh, almost ruled over the three districts of Nadia, Faridpur and 24 Parganahs, and was a terror to the planters. His headquarters were in Narkulbariah, in 24 Parganahs. His followers, '... affecting a kind of military order, marching in ranks under his lieutenant Ghulam Masud ...' attacked several indigo factories. They plundered the factory at Bhaduria, and took prisoner the manager of the Hugli factory with his whole family, and brought them before Titu Mir; they were released only after the manager '... wisely assented and agreed to sow indigo for them as rulers of India'. In 1832 indigo peasants, who had not been paid for two years, locked up planter Andrews of Khalbolia factory in Nadia in his own house, and threatened to kill him if they were not paid (RIC 1860: ¶1278 Rev. Blumhardt of CMS). Persistent representations from the planters and the *zamindars* bore fruit. A contingent of Calcutta Militia, under an English officer, was sent; after this was routed a regular military expedition arrived with artillery. In the ensuing pitched battle Titu Mir fell; Ghulam Masud was executed and the rest imprisoned.

Barasat thus had a history of militancy, and became prominent again during the indigo uprising. As Kling (1977: 61) observes, 'though the movement was checked in 1859, many of the peasants were Farazis who participated in the indigo disturbances—skilled in military organisation and the use of arms'. Defiance was in the air. The missionaries, in their evidence before the Indigo Commission, attributed the growth of '... independent habits ...' to the wide repercussions of the Santhal uprising (1856–7) and the Mutiny of 1857 (Chowdhury 1964: 201). The Santhals had combined under the leadership of the two brothers Sadoo and Kanoo to fight the oppression of money-lenders, *zamindars*, indigo-planters and the British Raj. 'Armed with bows and poisoned arrows, axes, swords and few guns, they carried all before them for a time through the western district' (Buckland 1901: I, 12–13). The British lost one sergeant-major, twenty-five men of the corps and an indigo-planter in an encounter; there was fighting at the border of Murshidabad, near the Kudumsar Indigo Concern of Maseyk—the man who was reported to have kidnapped the *zamindar* Jaggobandha Dutt (cf. p. 65). Interestingly, it was the villagers of Kalinagar, near the town of Murshidabad who refused to take advances from Watson & Co. in the autumn of 1859 and resisted the Company's armed retainers with their sticks (Kling 1977: 78). Another example was that of Gorai Mondal who, in 1854, together with some others, fought back the planter of a Jessore factory, when he attempted to get indigo sown by force. When he took the matter to the *nizamat adalat*, the peasants were set free, '... giving them credit for having only asserted the natural right of defence of their own property, which the aggressive acts of their assailants had encroached upon' (RIC 1860: Appendix ¶21). It is not improbable that, encouraged by the judgement of the *nizamat adalat*, the peasants of Barasat asked Mr Abdul Latif, Deputy Magistrate of Kalaroa, for protection against Mackenzie, the Manager of the Jingergattcha Concern, who they feared, had planned to occupy their lands forcibly. Abdul Latif issued an order to the planter, forbidding him to carry out his plan. Those were the days of the pro-planter Lieutenant-Governor, Halliday, and the number of complaints which had already reached him from the planters against Latif were numerous. Briefly mentioned, these were

i) his refusal to comply with the request of the planter to name the ring-leaders who agitated against the factory;
ii) telling the planters to go to court if they had grievances against the peasant;
iii) stopping the sowing of indigo whenever a peasant sent him a petition;
iv) despatching the police to prevent factory-servants from entering villages.

These were serious offences, but Abdul Latif's behaviour was unpardonably offensive to the planter, Mackenzie: 'He had been disrespectful in his form of address towards the planter' (Kling 1977: 69). Remembering the time when '... to the great unwashed abroad [Indians were] simply niggers, without a past, perhaps without a future ...' (Girish Chandra Ghosh in Moorhouse 1984: 76), Latif's crime was indeed grave, and Halliday took steps to put him in his place. He was transferred to a '... less important and sensitive area' (Kling 1977: 69). But what transpired from the complaint of the proprietor of the factory was of importance in the context of the movements of resistance. He said:

> ... that though never before had there been any complaints in any of the concerns' ninety villages, no sooner does Abdul Latif obtain power, than the *ryots* in the whole of that part of the country refuse to sow and fulfill their contracts (Kling 1977: 69).

Those who had been dumb all those years were now beginning to find their voice. It was again in Barasat, in 1855, that the peasants renounced indigo cultivation of the Barasat concern of the Bengal Indigo Company, putting them to a loss. J.H. Mangles, another outstanding officer, like Abdul Latif, had ruled in a dispute that the concern had forced the peasants to take advance. For giving this ruling Mangles was reprimanded by Halliday himself, and when in his report of the indigo dispute Mangles condemned the indigo system, he was transferred.

Things had begun to happen, and it was in Barasat once again, in 1856, that the peasants refused to sow indigo for the Habra factory; they explained to the Magistrate that they found tobacco and other crops far more profitable. The factory servants pillaged their houses, and there was a big clash between the Habra factory and the peasants (Chowdhury 1964: 198). That there was change in the old, cringing attitude of the common people towards authority had been noticed by many; Alexander Duff found, for example, that 'people were generally less tame' than before, and *The Hindoo Patriot* commented editorially (13 July 1854) on '... the growing estrangement between the body of the Nation (peasants) and their natural heads (the landlords)'. The illiterate wife of a tenant of Jaykrishna Mukherjee had gone to court, accusing the big *zamindar* of kidnapping her husband; finding him guilty, the District Magistrate of Howrah had sentenced him to four months rigorous imprisonment, or two hundred rupees fine, in lieu of hard labour (Mukherjee 1975: 209–10). Even if this was a charge trumped-up by discontented peasants, it was '... an index of the Native mind', as Reverend Long remarked. City wags composed verses commemorating the event, and on the occasion of the major festival, Durga-Puja, poets (*kavi-walas*) sang lampoons about Jaykrishna's case (Mukherjee 1975: 210–11). To many, Jaykrishna Mukherjee symbolized *zamindari* oppression, and people were now less prepared to tolerate oppression from any quarters.

A wind of change had begun to blow, bringing in some visible improvement in the economic life of a section of the peasantry. Cultivators in Barasat had changed over from rice to jute, finding this cash crop immensely profitable, as the demand for gunnybags was increasing due to the Crimean War (1854–7). The influx of troops in near-by Dum Dum and Barrackpore made the sale of local products easier and more lucrative; the army paid well, and seldom went into the matter of prices. To meet their demand, some cultivators took to growing only potatoes, instead of rice. With their economic improvement, peasants were becoming politically conscious. The proximity of Calcutta, spread of modern education, and the proliferation of vernacular newspapers—rising from 8000 copies in 1826 to 600,000 in 1857—had, to some extent, contributed to this new awareness. As Reverend Long observed, 'Indigo-planting had been the subject for the last sixteen years of incessant attack in their newspapers and the opinion of these papers filtered down to the masses.' A pamphlet entitled *Oppression of the Indigo Planters* was being distributed freely by Durgadas Chatterjee of Uttarpara (Mukherjee 1975: 235). Through circulation of pamphlets, books, newspapers and frequent visits of journalists in the villages—anathema to *Neel Darpan's* Planter Wood—'... new ideas had been slowly but surely beginning to glimmer on the darkened understanding of many' (RIC 1860: Duff, Appendix Part 3). Songs, some full of scorn, some sad and melancholy were set to tune. Peasants heard their own innermost feelings expressed in them, and feelings of discontent, suppressed for centuries, became intensified.

There was a new atmosphere of sympathy conveyed to the peasants through various channels; nobility and gentry became their champions. Among them were big *zamindars* working through their agents, small land-holders, most of whom were hostile to the trespassing of planters into their reserved territories of grain-dealing and money-lending; disgruntled *mondals,* dismissed employees of planters, and above all the missionaries, who lived in the villages near the peasants. It is not improbable, however, that it was the comparative economic prosperity of their friends and neighbours which had the biggest impact on their drooping spirits. The indigo-cultivators began to perceive that the phenomenal rise in the prices of all consumer goods, which they thought was a curse, and the cause of so much misery, could in fact be a great boon, a source of plenty—provided they had the freedom to cultivate the crop of their choice. They sought this freedom, by resolving not to sow indigo.

Barasat had virtually become a storm-centre—protests, refusals to cultivate indigo, to take advances, as also threats of forcible cultivation were frequent. In February 1859, when the peasants again refused to take advance from the Habra Concern, they noticed that the new magistrate, Ashley Eden, who had succeeded J.P. Mangles, was no ally of the Concern. Not only did he refuse to help the factory by pressurizing the peasants, but asked it to pay them a better price. The news spread, and there were petitions from peasants of the Barasat concern of Larmour, asking for protection against forced cultivation. Help was not long in coming; it came in the issuance of an order, forthright and unequivocal, to the Deputy Magistrate of Mitterhaut on 16 March 1859. This was the first time any magistrate had recognized the right of the peasant to sow his lands with whatever crops he liked; no one could by violence and without his consent sow any other crop. The order was to

> ... send policemen to the *ryots'* lands, to prevent any disturbances that are likely to ensue from any compulsory cultivation of their lands, and instruct them not to allow any one to interfere with that, if the land is really that of the *ryots'*. If the *ryots* wish to sow indigo, or anything else, the policemen will see that there is no disturbance (Kling 1977: 71).

The order made the planters furious; it was as if a pebble had been thrown into a hornest's nest. The secretary of the Planters' Association, William Theobald, observed: 'For a Magistrate to encourage what is at once contrary to law and morality, and to the fair interest of the capital, is intolerable' (*Indian Field*, 26 November 1859, in Ryot 1860: 8). Larmour, the powerful manager of the Bengal Indigo Company, was already angry; Eden had not only refused to modify the order—though he had personally written to him to do so—but had the affrontry to write and say that his letter should have been sent in the form of a petition, on a stamped paper, adding,

... I do not consider it desirable to carry on a correspondence with a party to a suit in my court on subjects connected with a case (Kling 1977: 70–2).

Planters had long been used to dealing with magistrates like Moloney, Skinner, Taylor, and Maclean—all good friends, always ready to oblige. The sudden emergence of a magistrate like Eden, with his innate sense of justice, his correct behaviour, and his capacity to see the peasants' problems, appeared to these planters as acts of unjustifiable hostility. Larmour complained to the Commissioner of Nadia, Grote, that the '... Joint Magistrate had done all in his power to instigate the *ryots* not to sow' (Kling 1977: 159). The Commissioner found Eden's order objectionable on the ground that '... it imposed upon the police-inspector (*darogah*) undue responsibility'. It was not for the police to decide the ownership of the land; they should rather be ordered to warn the peasants against disturbing, and not give them protection. John Peter Grant, the second Lieutenant-Governor of Bengal, had taken office just as this controversy was developing; the file was sent to him for his decision.

The Lieutenant-Governor gave it as his opinion that Sir A. Eden's principle was a true exposition of the law as it stood, according to which the police were bound to protect persons and property from unlawful violence, and to abstain from disputes respecting alleged contracts which were only cognizable by civil courts (Buckland 1901: I, 185).

This was the first decision of Grant on indigo matters after he took office, and his ruling was that 'indigo can not be supported at the expense of justice'. On 17 August Eden sent a letter, based on his own statement made in February, to the three Deputy Magistrates of Barasat for their information.

A New Lieutenant-Governor

John Peter Grant was no stranger to the people of Bengal. As Secretary of Lord Dalhousie, the whole weight of the government of Bengal had virtually rested on his shoulders, and he had implemented several administrative reforms. He had his own idea of justice, and held the view that it was the duty of the Government to do '... what the people themselves would do if they had the necessary organisation and resources' (Seton-Karr 1899: 120). He had great faith in the efficacy of law, and believed, as the distinguished *sadar*-court judge Alexander Ross was to do, in the 1930s, that 'law was to be the sole instrument of change'. As already mentioned, to bring colonial law and justice nearer to the villages, Grant had increased the number of subdivisions and courts in the two chief indigo districts of Nadia. His idea was to give the peasants access to the magistrates' courts, to show them that through the '... improved administration ... practically the protection of the law was no longer hopeless' (Buckland 1901: I, 145). He wrote strongly against the employment of compulsory labour in the irrigation works in Madras (Seton-Karr 1899: 124), as he did again in 1860 about the exploitation of·indigo cultivators. After thirty-one years of service in various departments of the Governments of India and Bengal, he was appointed, on 1 May 1859, to succeed Sir Fredrick Halliday as second Lieutenant-Governor of Bengal. Welcoming his appointment, *The Hindoo Patriot* paid him a rich tribute, and wrote in its editorial:

> He has given ample evidences of a vigorous understanding, strong will, independence of character and thorough appreciation of the duties of an Indian statesman.

The paper (12 May 1859) assured the new Lieutenant-Governor that '... his stern uprightness and brilliant intellectual abilities will go far to instill public confidence in him'.

Grant left on his first official tour in early August. While returning, his boat touched Krishnagar and Berhampore; villagers from Nadia had followed him from Krishnagar, and at Berhampore they handed him their petitions, complaining of the oppression of the planter William White of the Bansberia Concern. Grant took their petitions and ordered an enquiry by the Commissioner of Nadia. The investigation revealed that the magistrate had sided with the planter, and had dismissed the villagers' accusation of plunder as not proved. Of the two men whom the planter had had kidnapped, one was missing and was later discovered to have died while in the planter's lock-up. The other kidnapped man had once been an employee of the planter, and had been flogged and confined because he had refused to accept advances; he had been forced to settle the case out of court. This man was released and his case was taken up for hearing in the court. It was a sensation the way the petitions had been accepted by the Lieutenant-Governor and the peoples' grievances removed. It was reported in detail in the press and the contrast between the two Lieutenant-Governors in their behaviour towards the public was pointed out. Halliday also had been given a petition, asking for help against the oppression of planters; those petitioners had been influential citizens, not humble villagers, as in the present case. Halliday too had been returning by boat from his first official tour, but he had rejected the petitions as 'vague and probably not credible'—'they were never unfolded', commented *The Indian Field* (10 December 1859). Now, after all these years, the peasants were astonished at getting justice done. Another gruesome case came to light, where a peasant, wounded in an affray by factory servants, had died while being carried from factory to factory. Numerous petitions complaining of oppression now began to pour in, and most of them were found to be true. In his evidence before the Indigo Commission on 5 January 1860, one Sheikh Khosal Mundol, for example, stated:

> The factory of Ruttonpore is two or three *coss* distant from my village. It belongs to Mr Goldsmith, who is under Mr Furlong, who stays at Nichindipore. Mr Goldsmith has committed much oppression on the *ryots* of my village. On the 13th of last *Kartick*, by Mr Goldsmith's orders, Judoo Roy, the *Dewan* of Huttra Factory, seized Chunder Biswas and Buroo Mundol and Anundo Biswas and Sheikh Jumeeut; the last of whom made his escape from Ruttonpore Factory. The other three were taken away and since then we have seen or heard nothing of them. The reason why these men were seized was, that the indigo planter wished them to sow indigo by force which they refused to do. Then the indigo planters' people, to compel them and us who refused to sow indigo did all sort of oppression, looted us, and placed elephants and *lattials* from the factory in various places, so as to damage our crops and ourselves. ... We petitioned the magistrate but received no advantage from it. The magistrate was a friend of all the planters and would not listen to us poor men. ... We got no justice; and whether the three men are dead or

alive I know not. There are fourteen of us who have fled from our village ... because we are so oppressed by the indigo planter....

The general picture that emerged, deplored Sir Bartle Frere (1861), was '... a grievous reproach to any civilised Government. ...' and he found it outrageous that '... the Province, one of the largest and most easily governed of any we possess in India ... should have been subject to such a system ...'. Steps were taken to strengthen administration; officers were instructed to observe strict neutrality in all matters concerning indigo. Such instructions of the Lieutenant-Governor, however, were not easy to follow; as Sir Bartle remarked,

... high officials under Government were supposed to be deeply interested in indigo, and the *ryots'* belief that the Government wished them to sow indigo was something more than a misapprehension....

This belief was widespread among the villagers and Dinabandhu Mitra refers to this in his play; one of the peasants suggests that the Governor may have a share in the indigo business (Act II, Sc. 1). After the enactment of Act XI, 1860 it was the members of the '... uncovenanted judicial service of Government ...' whose '... lusters shamed the brightest brightness of the covenanted branch of the same service'; Herschel was '... the only one out of all the covenanted Magistrates of Jessore and Nadia empowered to adjudicate cases between the planters and *ryots* ...' who showed '... moral courage and integrity. ...' (*The Hindoo Patriot*, 11 July 1860).

Meanwhile, however, the letter sent by Eden to his three officers in Barasat was circulated, without his knowledge, by the Deputy Magistrate of Kalaroa as a *parwana*; this happened just a few days after Grant returned from his tour. The rumour now spread that the Government was actually opposed to indigo plantation. Grant immediately issued a notification to the effect that

... *raiyats* who had contracted to cultivate indigo must expect to be forced to fulfill their obligations; but no *raiyat* was forced to contract to cultivate who did not choose to do so (Buckland 1901: I, 186).

The notification served no purpose, and if anything, made matters worse. Planters insisting on the first part, claiming their pound of flesh, and peasants on the second, claiming that the Government was on their side. As mentioned earlier, Larmour had already complained, in April 1859, '... of a general disinclination among the *raiyats* of his concern to cultivate indigo', and had blamed the District Magistrate for his '... open hostility to the interests of the indigo planting' (Buckland 1901: I, 186). But Halliday, who had given '... a good deal of his attention to the complaints of the planters ...' (Kling 1977: 66), was no longer in office and no redress came. On the contrary, the decision of the District Magistrate was endorsed by the new Lieutenant-Governor, and police was instructed to protect the peasants. After the circulation of Grant's notification, Mssrs. Robert Watson & Co., Indigo Planters and Silk Manufacturers, wrote to the Government of India requesting '... His Lordship in Council to intervene to remedy the mischief caused by the Government of Bengal' (Watson 1859). It was an angry letter, full of resentment against the Lieutenant-Governor's insistence on

> ... the proof required by law—(and) ... no district officer can venture to accept less—of the conspiracies and illegalities from which we are daily suffering, of which convincing circumstantial evidence are available.

The Governor-General merely forwarded the letter to his Lieutenant-Governor, and Seton-Karr, Secretary to the Government of Bengal, sent a reply. He wrote:

> ... even if such 'general combination' and 'mutinuous spirit' as is assumed by Mr Cockburn, manager of the indigo concern Watson & Co., is ... contrary to all present appearances, to be found really to exist, the Lieutenant-Governor would look into the remedy, not for vigour beyond law to be exercised against one of the parties, without requiring technical proof such as Mr Cockburn in general terms apparently recommends (Seton-Karr 1860).

The tone of the reply was sufficiently sarcastic, so the matter rested there till the *parwana* referred to above was circulated. This time, a long letter came to the Government of Bengal itself, from the indigo planters Mssrs. Robert Watson & Co. in the districts of 'Moorshidabad, Nuddia, Rajshahye and Pubna'. The letter said that, '... notwithstanding excitement prevailing in other quarters ...', their concern remained unaffected till '... the Vernacular Proclamation of the Lieutenant-Governor was circulated through our villages in the Shikarpore Concern by the Kureempore Police authorities,' and their peasants were informed in '... adjoining Fureedpore and Baoromarriah, Rajshahye and Pubna by means of copies of the Proclamation from Krishnaghur and also from Shikarpore, where most of the head-*ryots* hold lands.' Throughout these areas the peasants had stopped cultivating indigo, and were determined not to fulfill their contracts. The letter continued:

> Even the more favourably disposed state that as Government has ordered them, by Proclamation, circulated from village to village, not to sow they can not disobey that order till another Proclamation is circulated ordering them to sow.

Clearly, these 'good ones' were afraid to go against the '... more evil-disposed—by far the greater number ...'—who were determined,

> ... even in the presence of the authorities, to resort to force and bloodshed rather than sow any longer, and our cultivation lands (in most cases our private property) are being daily wrested from us forcibly.

The peasants, who had absorbed the lessons of forcible occupation of lands from their masters, had begun to put them into practice. Much more serious for the planters was the virtual strike by their factory and household servants.

> In many cases they are compelled to relinquish their employment and join the conspiracy, supply of food and other necessaries are refused them and they are subjected to numerous other outrages.

Apparently, 'the head-*ryots* of Shikarpore and Krishnaghur'—whoever they might have been—knew how to plan and organize the uprising. They were consolidating their ranks with men and women belonging to their own class and castes, who worked as servants in the planters' factories and homes. In village society domestic servants employed by Europeans were either low-status Muslims or low-caste Hindus—*goalas, pods, doms*—some of whom might have been converted by the missionaries; they never belonged to any of the upper castes. For Europeans used to a certain style of living, to be without servants in a *moffassal* could be a disaster, so the weapon used was to deprive them of such labour. Whether it was voluntary absence from duty or not, those who worked as servants stayed back, rather than face violence A case history quoted by Ranajit Guha illustrates what could happen to those who refused to conform:

On May 11, 1858 about eighty inhabitants of the village of Betai (Nadia District) attacked Mr A. Hill's *cutcherry* in that village, took several papers, wounded a Government *peada* and *chowkeydar*, then went to another quarter of the village and plundered there and severely wounded six men. This last attack was made, because the sufferers would not join them against the factory (Case No. 18, 1858, IRC, App. II in Guha 1983: 201).

Another representation was sent by another planter from Nadia to the Commissioner on 20 February 1859, saying that '... mischief had been done in his indigo factories by the issue of a *parwana* from the Deputy Magistrate's court in Kalaroa'. But the Government found that ' the *parwana* had no influence in Nadia, where no one probably ever knew of it' (Buckland 1901: I, 186). This, however, was far from true. As Eden testified before the Indigo Commission,

Ryots came from Jessore and Krishnaghur and took authenticated copies of my order, knowing that the effect of the intimation would be to spread gradually throughout Bengal a knowledge of the fact that it was optional with the *ryots* to enter into contracts or not, as they thought fit (Kling 1977: 75).

Since no action was taken, the peasants felt convinced that the Government was with them and Torap, in *Neel Darpan,* could not have been the only one who prayed to God to grant the Lieutenant-Governor a long life: '... we shall have food to eat and the indigo devil of the bastards won't be sitting on our backs' (Act II, Sc. 1)

DEFIANCE

A few weeks before the spring sowing was to begin, the Aurangabad Concern in the northwest corner of Murshidabad was rocked by violence. With a new spirit of defiance the peasants surged forward to attack the Ancoor and Kalapani factories of David Andrews, where oppression and extortion had reached beyond the limits of their endurance. The planter had ordered an increase in the production of indigo by several hundred *bighas* in his *ilaqa* cultivation; since he was their landlord, the peasants had to carry out his oppressive demand. The *gomastha* of the factory, one Mir Tuffazil Hussain, whose appetite for bribes was insatiable, asked to be paid each time a peasant went to the bathing *ghat*; the only path leading there had been taken over by him to sow indigo. They had to pay him to prevent their rice-crops from being ploughed up, but he overshot himself when he threatened to force the women in a Muslim village to weed indigo unless he was paid eighty rupees. Not satisfied with half this sum, he took away some of their men. The furious villagers first rescued their friends and then swore revenge. When Mir visited this village next he galloped back to the factory sensing some change in the mood of the villagers; hundreds of them gave him a chase, and pulling him out of the shed in which he had hid, thrashed him mercilessly. By the end of the day three thousand peasants had assembled to march on to the factory. The frightened police merely looked on. According to plan, five hundred men gathered near the factory next morning to the beat of drums, to drag out the wife and daughter of the Assistant Manager of the factory and force them to weed indigo. But luckily for these women they had fled, narrowly escaping the fury of the people. Their next target of attack was the Bakrabad factory, another concern of Andrews across the Ganges, in Malda. They plundered the factory, destroyed the ledgers and took away several guns and a sword, of which planters kept a big store, as a *darogah* was to testify before the Indigo Commission. They used these weapons in their several violent affrays and while forcibly occupying the lands of the villagers. The next factory to be attacked was the nearby Beniagaon factory of one Mr Lyons. The planter opened fire and killed one and wounded five of the peasants; eventually three peasant leaders, Murad Biswas, Sauhaus Biswas and Lall Chand Saha—minor landholders who had organized the rebels—were arrested and tried (for details see Kling 1977: 91–3), but the peasants did abandon the cultivation of indigo. Another incident of revolt which occurred in the winter of 1860, in the Ratnapur factory, soon after the attack on the Auranagabad concern, was so unique that it should be narrated in detail. Two years previously, in 1858, to save their lands from being leased out or sold off to a planter by the *zamindar*, the villagers had gone to the nearby mission and pleaded with Reverend Bomwetsch to take the lease of these lands. The peasants had promised to raise half the purchase price themselves, but Bomwetsch could not help them. Feeling frustrated and helpless, they agreed to sow indigo for the factory under the threat of their village being destroyed by the planter. In the two years since then a change had come over these villagers; they had learnt to rely on their own combined strength, and drawing on their own spirit of defiance, the cultivators of the Ratnapur factory now refused to take advance and sign the contract. It was not so much to protest against the oppression of the management, as to recover their own lost dignity. After their refusal, hundreds of *lathiyals* were sent by the factory to surround their village, Bilabhpur.

> The villagers had organised themselves into 'companies', each specializing in the weapon it found most suitable—bows and arrows, slings ... bricks, bale-fruit, brass plates thrown horizontally, earthen pots thrown by the women, *lathis*, and finally, the fiercest of all, a dozen men armed with spears (Bomwetsch, 11 February 1860, in Kling 1977: 102).

The *lathiyals* were ultimately beaten back by the peasants who from then on did not sow indigo anymore.

After this, for a while, all looked quiet—like a lull before the approaching storm. The disturbances till now had been symbolic of the passionate anger of the peasants against the tyranny and extortion of some particular concerns located in some specific areas of Lower Bengal. The general revolt was yet to come, '... as a large surging wave comes, drawing its force from a countless number of small waves which grow into a large one ultimately to collide impetuously against the shore' (Chowdhury 1964: 165). Preparations for this were, however, already afoot, as the special investigator, Brownewood, reported:

A regular league was now formed against indigo cultivation, oaths were subscribed to by both Hindoos and Mussulmans, *ryots* of one village were called upon, by beat of drum, to assist those of another, if molested by the planters' servants, etc., and if pressed to cultivate indigo by such servants they were to resist; signals were made and given, subscriptions raised; villagers turned out by the beat of drum and proceeded in large bodies armed to any alleged threatened spot; in fact they had it all their own way, the police were afraid and had been bought over by the *ryots* (Kling 1977: 93).

News of this organized protest reached the Lieutenant-Governor when he returned from his tour of Bihar, on 10 March. Fearing that it would spread, Grant at once despatched the military police to Nadia. A few days later an important delegation from the Indigo Planters' Association including the secretary of the Bengal Indigo Company and the editor of *The Bengal Hurkaru* met the Lieutenant-Governor. The association attributed the trouble to '... a mistaken belief as to the views of Government in regard to the cultivation of indigo' (Buckland 1901: I, 86). Although the *parwana* issued by Eden was not mentioned in so many words during the discussions, it had never ceased to rankle the planters. When asked by the Indigo Commission to explain '... the reasons of this extensive unwillingness to sow...' Edward Chardon of Goldar factory, Nadia, replied:

There were two or three reasons, the principal one was the *parwanna* issued by Mr Eden, the contents were known to my *ryots* before the October sowings, I was surprised to find that this news had flown so fast among my *ryots* (RIC 1860: ¶2781)

Another planter, Durup de Dombel of Krishnagar, also felt the same way, and said:

From frequent conversations with all classes of men, I have been led to believe that a *parwanna* from Mr Eden, the Magistrate of Baraset, by his under officials, as also owing to the dissatisfaction spread by certain missionaries ... the *ryots* have risen in a body with a hope of driving the planters from the district, by the help of Government (RIC 1860: ¶748).

The Planters' Association asked for two measures 'to protect their interest thus damaged'; first, a Government notification to remove the mistaken impression that it opposed indigo planting, and second, a special law to make the breach of an agreement to cultivate indigo punishable summarily by a magistrate (Buckland 1901; I, 186). The Lieutenant-Governor acceded to the first request, but was opposed to the 'principle of the special legislation' (Kling 1977: 126). He argued that 'it involved questions of a very serious nature' (Buckland 1901: I, 187). The following day he issued a notification correcting '... any erroneous ideas as to the wishes and policy of the Government and impressing upon the *raiyats* the necessity of fulfilling existing engagements' (Buckland 1901: I, 187). As in his earlier notification, the officials were told to explain to the peasants that it was always optional for them to enter into a contract. Four detachments of military police were also sent to the Commissioner of Nadia to quell any disturbances by the peasants on their hearing the notification (*The Hindoo Patriot*, 17 April 1860). It is interesting to note that Grant explained to the Commissioner in a covering letter that it was not intended that the

police should interfere for the purpose of enforcing civil contracts. Breach of contract ... can be remedied only in the manner authorised by law (Kling 1977: 126).

This letter of 14 March, merely reiterated Grant's old policy, without indicating that a special law was on the anvil and was to be introduced in less than two weeks as a Bill to 'enforce the fulfillment of Indigo Contracts.' The 'law' was going to be a summary one. On 31 March the Bill was passed with stringent amendments detrimental to the peasants; Grant presented Bengal with his Act XI of 1860.

After thirty years the planters got a summary law, before the spring sowing started in April. The present chief executive officer in Bengal was, unlike Bentinck, no supporter of planters; on the contrary, he had bluntly disapproved of their behaviour. As late as in June 1859 he had rejected a law proposed in their favour (Chowdhury 1964: 164), but now he himself passed a special law for them, the sole purpose of which was to coerce the peasant into sowing indigo. Even on the eve of the Bill's being passed, Grant admitted in a letter to his friend Sconce:

> In practice, the *ryot* is made to act like a slave, not like a free man. … The present struggle on the part of the *ryots* is to avoid the cultivation of indigo; from this it is certain that *ryots* who cultivate indigo are forced to do so by illegitimate coercion (Chowdhury 1964: I, 164).

But the 'combination among a mass of *raiyats*', which had 'threatened a commercial calamity', looked so ominous and alarming that perhaps Grant's inner convictions were surrendered to the colonial Government's larger interests of commerce and trade.

Under Act XI, a magistrate could summon a peasant if he delayed or neglected to cultivate the whole of the land according to agreement, after having taken a cash advance 'on the oath of a planter who had made the advance or by any person on his behalf.' If the complaint was established, the peasant would be made to pay damages five times the advance made and five times the value of the seed he might have been given. 'Failing the immediate payment of the damages assessed', he would be imprisoned for three months. There was provision of penalty for intimidation—a prison sentence with or without hard labour for six months, or a fine of two hundred rupees, or both. For destroying or damaging crops, the same penalty could be meted out, and if the fine was not paid he would serve a further period of six months in jail. There was no appeal against the decision of the magistrate. The duration of the Act was for six months, and it recommended the appointment of a commission of enquiry into the practice of indigo planting in Bengal.

Grant justified the Bill, arguing that if the peasants had accepted the season's advances they had no moral right, still less legal right to turn suddenly round upon the planters and with nothing to complain of now more than they had before to refuse to do what, up to this moment, they have led the planters to expect that they would do according to custom and agreement.

The Lieutenant-Governor assured the peasants that after the current indigo season was over, they would have the freedom to refuse advances. The six-month validity of the Bill was destined to cover one entire indigo production cycle—from sowing, weeding and harvesting to delivery. Grant's advice to his officers to exercise their powers 'in such a manner as to do full and substantial justice to all parties' (cf. Kling 1977: 134) went almost entirely unheeded. The summary law allowed equal credence to the word of the planter and his factory servants if they swore they had given advances to the peasants. The magistrates—always more than willing to oblige their planter-friends—readily accepted their statements on oath as evidence, and rejected the sworn denials of the peasants. In one case, factory servants, who had been engaged long after the contract was alleged to have been signed, were deemed as reliable witnesses, their evidence was accepted unquestioningly, and the peasant was fined. In another case, a peasant testified that, even when he signed before the magistrate and his signature differed from the one alleged to have been his on the contract, his evidence was rejected (cf. Kling 1977: 159, 160). As a friend of the planter George Mears, the magistrate Maclean issued a *parwana* prohibiting the sowing of rice on lands claimed by the factory to have been under indigo the year before (Kling 1977: 152); but for another *parwana* issued immediately thereafter by Herschel and countermanding his Assistant Magistrate's, the rice crop of the villagers would have been ploughed up. Fear of famine (cf. *The Hindoo Patriot,* 11 July 1860) because of a long and continuous drought had gripped the villagers, but that was of no concern to Maclean. These were the administrators of the special law, 'corrupt, bribe-eating magistrates who have prostituted themselves', commented *The Hindoo Patriot* (Kling 1977: 149).

Magistrates took special care to convict peasants under each one of the Sections of the special law. By 25 April, 279 men had been imprisoned in Nadia district (Kling 1977: 148). The prisons were full; the imprisoned peasants were treated brutally till they agreed to sow indigo. Villages were empty:

> How many innocent people are in jail today ... the pitiful sight of their families is heart rending. Empty pots on the ovens, grains parching in the yards, cattle tethered inside the cowsheds, fields lying half-tilled ... The rice fields ... full of weeds.... (Act III, Sc. 2)

Advised by the planter, Furlong, the deputy magistrate of Damurhuda arrested the lawyer (*mukhtar*) of the peasants, Titu Chakravarty, on a charge of instigating them to refuse to sow indigo, and sentenced him to six months imprisonment with a fine of two hundred rupees. This case was a sensation in Calcutta, and was taken up by the Press, and later the Lieutenant-Governor referred to it scathingly in his Minutes. The idea was to terrorize lawyers so that the peasants would not be defended. There were others—humbler men, as innocent as Titu Chakravarty—who were not released as he finally was, but served their sentences; to pay their fines they sold off their ploughs and cattle, and became paupers. A peasant interviewed by the Indigo Commission inside the Krishnagar jail said that there was no suit against him for breach of contract, 'as they have sown indigo on my lands by force.' But he had been convicted on the false evidence of

> two syce and two spearmen of the factory, because I had complained against them for beating up two of my friends. The charge against me was 'instigating *ryots*'. I pleaded my innocence in my defence, but it was not listened to or believed by the Magistrate (*The Hindoo Patriot*, 25 July 1860).

Under this Section unjust convictions had become glaring and through the story of the imprisonment of old Golak Chandra Basu, Mitra has shown how easy it was to accuse innocent persons and get them convicted with the help of a few bribed witnesses, if the magistrate was biased.

Damages which the magistrate could assess under another Section were far in excess of the customary charge of ten rupees per *bigha*. Astul Mondol of Kanai Nagar, Nischindipore Concern told the Indigo Commission from inside the jail that he had been fined one hundred and ten rupees as damage for five and a half *bighas*:

> I did not sow, as I understood it was optional for us to sow or not to sow; villagers told me that my cattle and property have been sold and realised only thirty two *rupees*, I lie in prison for the balance (RIC 1860: ¶3228).

There are several such examples of gross injustice documented by Kling (1977) in great detail. Administrators of Act XI were in collusion with the planters and exhibited their uninhibited and unabashed bias towards their friends. The special law in their hands had become a new instrument of oppression. The anonymous roving correspondents of *The Hindoo Patriot* who exposed Skinner, Moloney, Taylor and a few others, were not safe in the district, as they 'dogged their steps like a shadow, and overheard every word they spoke and any consultation they held with their planter-friends' (*The Hindoo Patriot*, 25 July 1860). Failing to detect any of them, however, Skinner pounced upon one Shishir Babu as a suspect. He was a teacher in the village, but actually the most competent correspondent of the newspaper. Skinner publicly declared '*usko dekhenge*' (we'll show him). He managed to get four men to give false evidence against Shishir Babu's father as an instigator, and deputed his *darogah* to prove the case, 'whether true or not' (Bagal 1953: 28).

On rare occasions, when magistrates were impartial—as were Herschel, Bell, Davidson and Lushington—planters were exposed, and forged documents, which sent hundreds to jail, came to light. A letter from Lushington (15 May 1860), the officiating Commissioner of Nadia to the Government of Bengal, who tried a breach of contract case under Act XI 1860, is revealing:

> Concerned factory is Peerpoore, one of many forming the large concern of Mssrs. Hill & Co. General Manager of the factory is Furlong; Assistant in charge of this particular factory is Mr Sanbolle, who appeared as the chief evidence for the Prosecution. The case was brought up with the most complete proofs which could be possibly required: there were the books of the factory, written *Kaboolyats* with direct evidence to their preparation at the time of making the advances and the payment of the

advances. And every one of these proofs, completely, entirely, and I may add disgracefully, broke down.

The reply from the Lieutenant-Governor's office is interesting:

The Lieutenant-Governor does not understand on what grounds Mr Herschel (who tried the case as District Magistrate) determined not to enforce the Penal Section of the Act against the plantiff; the person responsible in this case should be put on trial on the criminal side for endeavouring to support a false claim by forgery (Young 1860).

There was another case of violation of agreement after the peasant was supposed to have been given advances. Chardon, the assistant manager of Goldar Company of planter Mears, produced himself, his *muhuri* and manager as witnesses and gave evidence as to the time the advance had been paid and counter-signed. Chardon testified on oath that he himself had written the agreement (*patnee jit*) and signed it before the advance was paid; when cross-examined, the manager testified that he did not know when it was signed, but 'I asked the *muhuri*, who had written the *patneejit*, to go to Mr Chardon and get it signed.' The *muhuri*, when examined, testified that 'it was filled up at the time of payment of advance, and signed by Mr Chardon at about 7 or 8 next morning.' The case was dismissed by the *sadar amin*, Davidson, as a clear example of perjury (*The Hindoo Patriot*, 11 July 1860). The manager of Lokenathpur Concern G.R.J. Mears, informed the President of the Indigo Commission that

Since the principal *sadar ameens* have come, I have withdrawn all case of breach of contract throughout the whole of the factories as it was perfectly useless carrying them on. The proofs which we are enabled to bring forward are not approved of by these gentlemen, although they were considered sufficient by Mr Moloney and Mr Maclean (RIC 1860: ¶3368).

Mears had gained decrees from Maclean on production of factory ledgers, which were 'supported by the evidence of the factory servants' and often by his assistants. He admitted that contracts were not on stamp papers, these only being filled up when he 'apprehended trouble from *ryots*'. Following Mears, several other planters withdrew their cases, fearing they would be found out (RIC 1860: ¶3368).

Though the law played havoc with the lives of peasants, *The Hindoo Patriot*, while condemning the administrators of the law, found no fault with the law itself. The paper commented, in a vein similar to that of Mitra in *Neel Darpan*: 'The Government took the only course possible under the circumstances.' What were the circumstances? The dumb had found their voice and the 'lower order of men' were no longer looking up to their 'natural heads', with their old 'filial regards'. Faith in British justice, *lat-saheb* and the *sarkar* was so deep-rooted and pervasive among the English-educated Bengali public of those days, that even *The Indian Field,* more liberal in its views than *The Hindoo Patriot*, wrote editorially: 'any hardship Act XI would inflict upon the cultivators would be offset by the great benefit to them of a Commission of Enquiry' (Kling 1977: 135). The '*maharani hawa*' had begun to blow, and it effected all; the author of *Neel Darpan* was no exception, nor were the villagers. As the Great Queen Empress of India, the Maharani Victoria was the symbol of benign motherhood, full of love for her oppressed subjects, putting them on her lap like her own children and suckling them herself. A forlorn widow, she needed the love and veneration of all her people. 'Villagers worshipped in their homes her effigy on the silver Rupees' (Lawrence 1928: 239), and as late as 1930 her statue used to be worshipped in Ayodhya:

> ... Sitaram reports that people used to worship the statue of Queen Victoria. He describes how the police had to prevent the pilgrims from climbing the pedestal to offer flowers, etc. The reason for this is contained in a legend telling how Nal, one of Ram's generals, fell in love here with a local girl. Ram prophesied that this girl's children would become famous and that in the Kali-yuga they would rule the whole world. This is how Queen Victoria came to be incorporated in the Hindu pantheon (Van der Veer 1989: 14).

The symbol had a basis, not only in the villagers' tendency to venerate motherhood, but also in Queen Victoria's widely known non-racist attitudes:

'... natives and coloured races should be treated with kindness and affection, as brothers—not—as alas! Englishmen too often do—as totally different beings to ourselves' (Thompson 1989).

Little could have Victoria imagined how her English subjects were using her name to threaten Bengal peasants. Planters fully utilized the ignorant peasant's awe and reverence for their sovereign to their own benefit; a rumour was circulated that the Queen had purchased all indigo factories and that the planters were now her own managers, and that she would compel the peasants to sow indigo for her (*The Hindoo Patriot*, 11 July 1860). All kinds of rumours were, in fact, circulated in the rural areas to intimidate the peasants into sowing indigo. One such rumour was that

the cavalry would cut down and the Gurkhas shoot down every man who will be found ploughing his rice lands unless with a magistrate's orders, and bullocks will furnish beef to the *Sahibs*.

Such rumours and threats no doubt frightened the peasants. The spectre of their treasured and holy bullocks served up as beef no doubt haunted them—the terrifying tales told among the *sepoys* just before the Indian Mutiny must have still been fresh in their minds. Yet for all this, they stood firm, determined not to sow indigo again for the planters. With all threats failing and legal coercion no longer successful, the planters resorted to a much tried weapon—violence. Larmour, manager of the Bengal Indigo Company, sent his musclemen (*lathiyals*) to burn down the home of one Molamdi Biswas, destroy the paddy crops of Beshohoree, Kakooria and Ramkrishnapore villages and sow indigo in their fields (*The Hindoo Patriot*, 23 July 1860). Similarly, Mears of the Lokenathpur factory set fire to the houses of villagers who refused to sow indigo (Kling 1977: 150). The indigo districts of Lower Bengal were in turmoil. Feeling betrayed by the government, the peasants were in a bitter and challenging mood. When the *darogah* went to Narainpur village in Bangaon to read out and explain to them Act XI, they mocked at him, and said they would obey no such order. The angry *darogah* arrested four of them, but two hundred peasants, appearing from a neighbouring village, attacked him and rescued the four men. A similar fate awaited the Joint Magistrate the following night, when he had nineteen men arrested without warrant, simply because they had been pointed out by a factory servant (*jamadar*) as enemies of that factory. He was forced to flee, leaving his prisoners behind (Kling 1977: 151–2).

Discontent, which used to manifest itself sporadically through spontaneous and isolated rebellion against the oppressive management of particular indigo concerns, had grown with the years and now refused to subside. Intolerable deprivation, oppression, exploitation, years of suppressed anger of thousands in Bengal's indigo villages provided the basis for a social movement. Organized direction was offered to the peasant by various groups hostile to the planters, to effect change in the face of resistance. The revolt broke out with the rising expectation of the peasant of approaching freedom to cultivate his own lands and raise whatever crops he chose to. In his evidence before the Indigo Commission, the perceptive District Magistrate W.J. Herschel mentioned this change in the peasants:

> There appeared among the *ryots* a general sense of approaching freedom. They behaved as if about to be released from something very oppressive, and as if impatient of the slowness of the process (RIC 1860: ¶2819)

On 10 April 1860, soon after the passing of Act XI, the first violent resistance against the planting of indigo occurred in Peeraree, a small village near Neeshanpur indigo factory in the district of Pabna. In the ensuing confrontation between the Deputy Magistrate's party of 'about 20' armed police and the two-thousand-strong group of villagers armed with bows and arrows, swords and sticks, the police was repulsed. Two policemen were injured and captured by the villagers; as the *havildar* Sheebho Khan reported, 'it is not known whether they were killed' (Ryot 1860: 309). The Deputy Magistrate fell off his wounded horse, but was carried away to safety by his men. A guest of the planters, he had underestimated the strength of the peasants, but later, when surrounded by them, he was so utterly demoralized that he would not let his men fire on the crowd. To save his master's life, however, the *havildar* opened fire, while the magistrate fled to the safety of the planter's home. Alarmed, the Government took 'prompt and vigorous action' against the village '*lathiyals*', as Buckland called them (1901: I, 186), and despatched troops, and 'with a judicious display of force the violent outbreak was suppressed' (Buckland 1901: I, 186). According to the planter Edward Chardon it was 'Mr Herschel's police who were at the head of the whole of these *ryots*' (RIC 1860: ¶2781), but according to Deputy Magistrate Lingham's information, it was Ramrattan Roy's agent who was responsible for this attack (Kling 1977: 158). Roy, one of the biggest *zamindars* in Bengal

> enjoyed authority over a region extending from the shores of the Bay of Bengal to Dinajpur, and was a terror to the indigo planters, who feared his armies of club-men and spear-men (Mukherjee 1975: 244–5).

The agent who organized the peasants in the uprising was Mahesh Chandra Chatterjee; he might have been deputed by Roy, or he might have done it on his own initiative, for indigo planters were his *bête noire*. Well educated, and himself a wealthy landholder, with the 'notoriety', mentioned earlier, of being 'the most dangerous agitator in Nadia'—a qualification in the eyes of every rebel—he had considerable influence over the peasants. He held nightly meetings, showed *parwanas* hostile to indigo planting, and advised his audiences not to sow indigo. It was under his advice that the peasants of the Katchikatta concern complained to the magistrate against the management. To help the peasants in legal matters, he used to pay the fees of the lawyer Narayan Chandra Mukherjee (Kling 1977: 86–7).

There were reports that peasants had not only abandoned the sowing of indigo, but that they had sworn to avenge themselves on the notoriously cruel factory employees and servants, whose frightful oppression they had to bear meekly for years. One special target of vengeance was Deenanath Banerjea, the manager of the Baghdanga factory in Kumreeah village of Nadia District. The intensity of their anger took a terrible turn, and in a body they began by beating up the *coolies* who had been sent to sow indigo in a nearby village; one of these *coolies*, Madan, was later reported missing—he had probably been murdered and his body made away with. Then, shouting in unison 'We shall knock the brain out of him if we can find him', they proceeded to village Kumreeah, where the manager was holding *cutcherry*, and attacked the place. Severely beaten up, Deenanath Banerjea ultimately managed to escape, but his friend Rammohan Banerjee, who tried to save him, was killed. So violent an outbreak shocked the Government, but the cause for the animosity of the peasants of Kumreeah against the servants and employees of the Bengal Indigo Company's factories could not be ascertained.

Various methods were devised by the peasants to harrass the planters. In early July 1860, the Meerpore Concern in Rajshahi District, which held the lease of a farm—for twelve *annas* in six villages from one partner and four *annas* from another—was refused the renewal of the twelve-*annas* lease which had just expired. This was rather unusual, for no *zamindar* liked to lose the excessively high lease money he almost always charged a planter. The four-*annas* lease still had three years to run, but the small landholder suddenly turned round and complained to the Court that he had given the lease only under duress, since he had been confined in a lock-up, and that therefore he was now cancelling it. Both partners had obviously combined with the peasants to oust the factory from the leased land—in other words, from the village itself. The peasants had long stopped work and refused to pay rent, and had not allowed the factory *tehsildars* to enter their village. This was one of the rare examples of village landholders actively participating in the peasants' resistance movement (Reid 1860a). Once a planter lost the lease of his estate, he was left without his *zamindari* right to coerce the peasant to take advances and sow indigo. The only way then left open to him to recover the balance, or fight for the renewal of the lease, was to go to the Civil Court. One such case was that of one Mr White of the Hanskali Concern, in Nadia. He lost the lease of Hanskali and Govindpur, and the peasants took this opportunity and refused to sow; all he could do was to wait for the Civil Court to take its course and recover his balances for him (Chowdhury 1964: 167). However, the situation was never so desperate for a planter that he had to close down a factory, as the Deputy Collector of Jessore had feared. The planters must, nevertheless, have considered the situation serious enough to have complained about it directly to the Government of Bengal. Steps were taken immediately, and the sub-divisional officer, Taylor, was asked to take up proceedings under the Indigo Enforcement Act XI of 1860. Soon after, a report was received from the Commissioner of Rajshahi saying that peasants had been complaining that their indigo was going to be cut by force, and that work in several factories had been interrupted. The *coolies* and boatmen of the Mohanapur Factory had gone on strike for the second time, and *goalas* had stopped working as servants. The Hizeahat, Sonakundoo and Salgamuddeah factories were in trouble; complaints against them were that they had rice lands forcibly sown with indigo and had cut it without payment; they had also threatened to forcibly sow indigo the following October and then adjust receipts against the accounts of the peasants in such a manner that this would be the basis for all future claims (Reid 1860b). The oppression of the *naib* was also a major complaint, but he was partially taken care of by the peasants themselves; hundreds of them stood aggressively on the river bank, armed with sticks, waiting for his palanquin (*palki*) to pass by:

> If these causes of distrust are not removed within the next few days, there will be no alternative (unless the Planters prosecute under Act XI) but to allow the remaining plant to rot on the ground (Reid 1860b).

Another serious uprising was reported by the officiating District Magistrate of Pabna, Bainebridge, in the southern part of the district. Peasants of Mssrs. Robert Watson & Co., one of the biggest indigo manufacturers, had combined and were determined not to sow indigo; they had 'cut off connection with the factory altogether.' The planter's servants were intimidated and warned off the village, and his *coolies* were not allowed to work. When 'Mr Roberts attempted to cultivate with the help of some of his *coolies*, he was chased off.' Bainebridge feared that this combination would encourage the men to resist, and that all the factory servants would 'covertly court and provoke aggression as their best policy.' The peasants squatted on the *chor* lands, claimed by the factory as *nij*. It was the 'intervention of the authorities' that prevented a 'violent collusion'. Bainebridge asked for 'a huge staff' to meet the pressure (Bainebridge 1860).

Many of the peasants, suggests Kling (1977: 157), were trained club-men, and enjoyed the patronage of Ramrattan Roy and his agent Mahesh Chandra Chatterjee. Violent confrontation was a mere question of time. Bainebridge had expected that the planters would '... not forego next season's crop without a struggle and they would attempt to oust the *chor*-squatters and raise the rent of others' (Bainebridge 1860). The peasants on the other hand were determined not to move out, or pay rent. The District Magistrate's advice to these hundreds of excited men was to first go to Court with their 'proofs' and have their 'occupancy rights established' under Section IV of Act IV of 1847, and only then protest 'without violence or breaking the seeds', if they found the lands sown forcibly with indigo. Bainebridge's threat, that if the peasants were violent 'it was the duty of the *hakim* to punish and not to protect' was as infructuous as was his advice. To the peasant a *hakim* who 'protected' had ceased to exist; one Deputy Magistrate had within four days convicted all peasants involved in seventy-nine cases—the defendants had hardly been examined, and one Tituram Chakravarty, who had come to defend them, was tried and convicted in ten minutes (Chowdhury 1964: I, 183–4). Every illiterate villager knew that the new law was a travesty of justice, and that it had been passed only to terrorize them into submitting to the planters' demands for indigo. As the Reverend Bomwetsch told the Indigo Commission, the peasants had 'almost lost faith in Government'. This was corroborated by Reverend Long: 'The Act had created a feeling among the *raiyats* of bitter hostility towards the authorities' (J.P. Grant 1860). To a simple village woman, Rebati, it was some new order from the government, under which 'the *sahebs* of the *kuthi*' and the magistrate '... can join hands and send to jail anyone they want for six months' (Act I, Sc. 4). The authorities also knew that 'the law protected the planters to the verge of justice and in truth ... excepting that the Act was declared temporary, it was unfair to the *raiyats*' (GG 1860a). The peasants had the feeling that they were fighting an unjust and unequal war on more than one front. On one side was the planter, with his trained clubmen and mounted elephants, his loaded guns and instruments of torture which were exhibited before the Indigo Commission (cf. *The Hindoo Patriot*, 28 July 1860); on the other side was the cruel, vindictive law, with the biased administrator who sent them to jail and reduced them to penury by forcing on them unfair, excessive damages. Chander Ghose of Nadia's Changuri village preferred to stay in prison: 'I shall have something to eat', he said (RIC 1860: ¶3283); 'I have become a *fakir*', said another peasant, Astul Mandal of Kamalnagar Concern, Nischindipur (RIC 1860: ¶3228). Staring at them on a third front was gnawing, grinding poverty, and the utter ruin of their families, especially when they were thrown in prison. The following official table (Herschel 1860) gives some idea of the extent of misery:

CHARGE I.	BREACH OF CONTRACT (1314b)		
NUMBER OF PERSONS CHARGED 8654	PERSONS CONDEMNED TO PAY 1245	GROSS AMOUNT OF DAMAGES IMPOSED 60880–6–0½	AVERAGE AMOUNT OF *BIGHA* 14.7.10
	NUMBER OF PERSONS IMPRISONED IN DEFAULT OF PAYMENT 383		
	AMOUNT OF DAMAGES PAID BY *RYOT* 31225–7–0½	AMOUNT REALISED BY SALES 2387–14–11½	

CHARGE II.	INTIMIDATION	
NUMBER OF CASES	NUMBER OF PERSONS	
	CHARGED	CONVICTED
34	230	35

CHARGE III.	MALICIOUS DESTRUCTION OF INDIGO	
NUMBER OF CASES	NUMBER OF PERSONS	
	CHARGED	CONVICTED
54	402	136

Little wonder then, that there was the intensity of feeling which Grant perceived and recorded in his Minutes (1860: Indigo Papers, Vol. 3, ¶4):

The feelings of the *raiyats* against Indigo sowing is so strong in the Bengal Districts South of the Ganges that it is vain to expect Indigo to be sown in them as usual the next season. Such was the determined aversion of the *ryots* to Indigo planting at the moment, that none would sow this season even though they should be offered Rs. 20/ a *bigha* for so doing.

That the unsoundness of the prevailing system had been pushed to the extreme and dangerous length in some parts of Bengal, was admitted by the Governor-General himself. But when Grant said that 'the question Government should immediately deal with is not the cause of the feeling, but the feeling itself', he was merely avoiding the issue (GB 1860). Advocating a policy of pacification was like wrapping up a festering wound, but not treating it. That a 'sounder system' existed in the Benaras Division for example, where 'indigo planters work[ed] with the *raiyats* instead of against them' was admitted and much appreciated (cf. Grant in Buckland 1901: I, 259), but it was never emulated. In fact, 'oppression of indigo cultivators was not known to have occurred in Northern India' (CEHI 1984: II, 260–1), where local *malguzars* dealt with the cultivators and acted as contractors for the European planters and there was more distance between planter and cultivator than in Bengal. In the Madras Presidency the situation seems to have been still better:

... Indigo is almost entirely grown by the *Ryots* who have no dislike to its culture. The manufacture of the dye is also very largely conducted by the Natives on their own account.... When the demand for the plant is sure, the *Ryots* cultivate it extensively without contracts or advances. Many manufacture their own produce, and all are at full liberty to ... the *Ryots of this Presidency are themselves the 'Zamindars'*, the *Ryotwaree* tenure giving them absolute right to their lands in perpetuity ... they are consequently free to raise what crops they please, and are fully aware of their rights as well as able and ready to assert them (SIM 1860).

Another similar report about the situation in the Madras Presidency states:

This plant has been cultivated here by the Natives ever since the commencement of the century, at first it was used in a rude state as a dye for local purposes. When it became prized in the European market, however, a large number of Factories were built in the District and carried on under European superintendence for some years.

The Natives themselves, about the year 1830, discovering that the process was a simple one, built Factories for themselves and soon were enabled to beat the Europeans out of the market, there being now not a single Factory in the District of those formerly established under European superintendence.... It frequently happens ... that the grower of the Indigo is both the Manufacturer and the Shipper ... (Wood 1860).

These Reports were there for all to see and act upon. If the authorities had been really serious about changing the system in Bengal, which they themselves denounced as 'evil' and which they blamed for all the difficulties in the indigo-growing areas, they could well have done so; but they did not. In Bengal the indigo planter retained direct control over primary production. Behind this direct control lurked the uncontrolled desire to amass a fortune, 'to have the largest possible profits and this was made out of the appropriation of the peasants' surplus' (Chowdhury 1964: 124). The process of appropriation led inevitably to increasing pressure on the peasant, and to increasing ruthlessness, both physical and economic. As testified by peasant after peasant, the consequences were forged contracts, cheating in payments, forced advances, ploughing up the peasants' rice lands, forcible sowing of indigo, kidnapping, detention in lock-ups, and false measurement of lands. The 'ruinously unprofitable, compulsory nature of the cultivation' (Chowdhury 1964: 124) was revealed by the Indigo Commission, but those responsible for this state of affairs were exonerated of practically all blames excepting two: kidnapping and seizing of cattle. So the peasants turned to their own people in the villages—to the *mondals*, the minor landlords and moneylenders, the small merchants, on whom they habitually depended; under their leadership they continued the struggle against indigo.

Two brothers, Digambar Biswas and Bishnu Charan Biswas—still remembered in the 1940s as the 'Ram and Lakshman of Nadia'—became important peasant leaders. They were minor landlords and moneylenders in their own village of Poragachia, and had left the service of the notorious planter, William White of the Bansberia Concern of Nadia. It had been against White's oppression that the peasants had gone all the way to complain to Grant, soon after he had taken over. The two brothers had a large following in different parts of Nadia, and sent messages to various villages to abandon indigo work. When the peasants of Govindpur, near Hanskhali, stopped cultivating, White sent '... more than a hundred *lathiyals* mounted on elephants, to attack Govindpur. But they were beaten off by spearmen, hired by Digambar, who were aided by villagers' (Kling 1977: 96). To protect his own village, which was attacked several times, Digambar Biswas had engaged his own band of clubmen, and to keep the movement alive, he spent the 'enormous sum of Rs 17,000' (Kling 1977: 95–6) to help peasants pay off their debts to the planters. Digambar Biswas eventually died a poor man, his funds all exhausted. There were other men like Biswas, who had to suffer because they helped the peasants in their resistance. All the three Mullick brothers of the Talookdar family, who owned Jayrampore village, were harrased and arrested on the charge of instigating the peasantry not to sow indigo. One of these brothers, Girish Mullick, had been manager of the factory but had been dismissed on the charge of dacoity and was then illegally detained by the planter; when he was released by the local constable, the Magistrate, Skinner, a friend of the planter, fined the constable and transferred him. Finally, Girish Mullick was convicted on the charge of 'exciting the peoples' mind and giving bad counsel'. His two brothers were freed, and the eldest, Ramrattan Mullick, became known as 'the indigo-plant destroyer' and as 'Nana of Bengal' (Kling 1977: 96–7), after Nana Saheb, the last Peshwa of the Marathas and a major figure in the Mutiny of 1857. Brindaban Chandra Pal Sarkar of Sibnibas was another man whose home was a regular meeting place for peasants; this man, detested by Larmour (cf. Kling 1977: 90), counselled them to renounce indigo. Like Biswas, he also helped them financially, even paying their penal damages (Kling 1977: 91). There must have been several others, whose names have not been recorded. If half of those with the name Mondal who gave evidence before the Indigo Commission inside the Krishnagar jail were village headmen or their relatives, their participation in the movement must have been considerable. These were the men who had been wooed and then used by the planters in the early years when they had wanted to lease rural lands; after they managed to become planter-*zamindars*, their once benign faces gradually turned upon these *mondals* as the faces of oppressors. In *Neel Darpan* the two brothers of the *mondal* who had fallen from grace and left the village after they had been beaten up for refusal to sow indigo, had vowed never to return, even if they had to beg from door to door (Act I, Sc. 1). A similar, real-life case was that of Podie Mondal of Badbagee. He had been forced to take advances to cultivate indigo on nine *bighas* of his own land:

> The whole of the lands which I had brought under cultivation, has been measured for indigo, with the exception of two *bighas*. I am in straightened circumstances and can not grow rice, having no land left for it. If I borrow rice from *Mahajans,* I shall not be able to repay it hereafter and the result will be my ruin ... (KT 1267 BC).

Podie Mondal had no option but to leave his village. When the District Magistrate William J. Herschel was asked by the Indigo Commission if he knew of any head peasants with 'sufficient resolutions and knowledge' to agitate their own peasants and form 'inter-village combination', he replied that he could 'point out hundreds such'. Local leaders had thus sprung up in any number. Outside the rural area the unfailing support which kept up the spirit of resistance came from the Indian Press; but for the courageous correspondents, who travelled to the interior of Jessore and Nadia, crimes and atrocities committed by those ruthless agents of colonization would have remained hidden from the public. The editor of *The Hindoo Patriot*, Harish Chandra Mukherjee, raised the peasants' demands from the ordinary commercial question concerning a blue dye to the political level. One of his editorials reads:

> Bengal might well be proud of the peasantry.... Wanting power, wealth, political knowledge and even leadership the peasantry of Bengal have brought about a revolution inferior in magnitude and importance to none that has happened in the social history of any other country. They have battled with adversaries possessing some of the most formidable elements of power.... They have achieved a success of which the benefits will reach all orders and the most distant generations of our countrymen (19 May 1860).

The other group of persons, who raised the issue before the eyes of their own countrymen abroad, were the missionaries. They lived in the villages, but were not of them and had also entered Bengal as agents of colonization on a 'civilizing mission' to convert to Christianity, to 'improve the moral conditions of a people whose social condition is extremely low' (RIC 1860: Evidence of Rev. G.G. Cuthbert, p. 179). 'They tried to make colonial peoples into disciplined agriculturists, or workers and obedient subjects of a bureaucratic State' (Cooper and Stoler 1989: 609–10). If more civilizing produced 'a less distinctive category of being' (Cooper and Stoler 1989: 609), 'a good nobody, which is what a servant is' (Kincaid 1986), that was not their problem. The problem arose when there was a clash of ideas between them and the other agents of colonization, also settled in the villages, and whose mission was anything but to improve the minds and morals of the colonized. Finding the coercive relationship of the planters with the peasants intolerable, the missionaries changed their policy of improving the heathen. They deplored

that of all the obstacles to the spread of Christianity in India, one of the greatest is the irreligious conduct of my own countrymen. Thousands of Natives have said to me, 'We judge of the Christian religion by what your countrymen do, not by what they say, by the life, not by the doctrine' (Rev. Long in *The Calcutta Christian Observer*, August 1861, p. 91).

The contribution of the missionaries to the cause of the oppressed peasantry must now be discussed in some detail.

THE ROLE OF MISSIONARIES

To a villager all Europeans looked alike. They were all fair complexioned, they spoke the same language, had a similar life-style and religion. To a peasant in the indigo districts, all white men were 'indigo devils' (*'neel mamdo'*, cf. Act II, Sc. 1). The activities of the missionaries had been legalized by the Charter Act of 1813, and soon they started to fulfill their pledge '... to rear indigenous Christianity' (Long: 86, *The Calcutta Christian Observer* on the *Nil Darpan* Controversy). To an ordinary rice-eating villager of any caste, the indigo planter and the missionary looked the same. These men had deprived him of his two most precious possessions—his rice and and his caste. A humorous verse in *Neel Darpan* describes how a zealous missionary is looked upon by the villager:

Your caste is robbed by the *padri*, Your rice is robbed by the indigo-monkey (Act II, Sc. 1).

Some village poet, stringing a few words together, had also given vent to his feelings:

The enemy of the soil is indigo, The enemy of the labourer is idleness; So the enemy of caste is Padri Hill (RIC 1860: ¶1693).

The activities of the missionaries, however, did not remain confined to conversion. Living near the people, learning their language, they saw from close quarters the misery of the indigo cultivators, and spoke out through various channels at their command, that the system of indigo production in Bengal was a forced system ... stained with oppression and cruelty (Long 1856 May: ibid 262). They carried on their crusade against the planters with boldness and consistency through several long conferences, articles and letters to their Societies in London. Their deliberations set off a regular discussion in the public press. 'In fact the first full-fledged article concerning indigo came out in the issue of 24 April 1856, containing reference to the discussion of the missionaries on indigo in their conference in September 1855' (Chanda 1984: 53).

The planters soon realized who their foes were, and to harm the reputation of the missionaries they did not hesitate to spread canards, allegations that they were inciting the peasants against them. They even sent James Furlong, a senior indigo planter, to the Bishop of Calcutta, to seek his intervention on their behalf, but failed. Undaunted, they began their attacks through the columns of the pro-planter newspapers—*The Englishman* and *The Dacca News;* but with an amazing solidarity all the periodicals of the London Mission Society, the Baptist Mission Society, the Church Mission Society and the *Calcutta Christian Observer* of the Calcutta Missionary Conference met the offensive launched by the planters (Chanda 1984: 48).

When the indigo revolt began, and there were reports of violence, the planters issued a virulent statement in the *Indigo Price Current* (8 March 1860):

For sometime the *ryots* in many of the largest indigo concerns in Bengal have envinced a disposition to break their contracts, and refuse to sow indigo, having been instigated thereto by evil disposed persons and aided by ... the Missionaries.

The Indigo Commission, constituted in May 1860, could not simply brush aside this serious charge, especially since there was a real crisis in the production of the dye. An enquiry was made into the specific allegation that the 'conduct of the missionaries' was one of the causes for the 'late crisis' (Chanda 1984: 49-51). After a thorough investigation, the missionaries were exonerated of all charges, and the Commission issued a report which speaks for itself:

A great deal of indignation has been evinced at Reverend gentlemen ... taking on themselves the character of political agitators; to express dislike of what they deem oppression, when forced on their notice and to stand up for the rights of those who have had no tongue to plead for them, be to carry on agitation, the Missionaries ... have done this. But in so doing they had no private interests to advance, and no political object to gain except the contentment and well being of the agricultural population (RIC 1860:¶120, 121).

Lieutenant Governor Grant also expressed his admiration for these 'excellent men' and felt that

it would have been inexcusable had they turned a deaf ear to the well-founded complaints of the people in close and kindly intercourse with whom they were living and some of whom belong to their flocks (Buckland 1901: I, 255).

There was yet another witness before the Commission—E.W.L. Tower, a former District Magistrate of Faridpur—who exonerated the missionaries of the alleged charge of having stated that 'Not a chest of indigo reached England, without being stained by human blood'. Tower claimed to have made this statement himself and went on to describe what he had observed in his District:

> I wish to state that considerable odium has been thrown on the missionaries for saying that, 'not a chest of indigo reached England without being stained with human blood'; that has been stated to be an anecdote. That expression is mine, and I adopt it in the fullest and broadest sense of its meaning as the result of my experience as Magistrate in the Faridpore District. I have seen several *raiyots* sent in to me as a Magistrate, who have been speared through the body. I have had *raiyots* before me who have been shot down by Mr Forde (a planter). I have put in record how others have been first speared and then kidnapped; and such a system of carrying on indigo, I consider to be a system of bloodshed (*The Hindoo Patriot* in Majumdar 1963: IX, Part I 921).

But for all the high praises showered on the missionaries, the Indigo Commission was reluctant to accept some of the serious charges brought by them in their evidence against the planters. Regarding the burning of houses for example, the Rev. Bomwetsch of the Church Missionary Society had stated that

> ... some people stated to me that some houses ... had been burnt down, because the people were reluctant to sow indigo, and in order to oblige the *zamindar* of the place, by harassing his people to give up his village to the planter.

After investigation the magistrate had in fact punished both parties, but the Indigo Commission rejected the verdict, as 'no well-proved instance of this sort ha(d) been brought to (their) notice in any oral evidence'. That a missionary 'had known of such acts' was not considered proof enough, although the concerned magistrate had in this particular case investigated the matter and found the parties guilty (RIC 1860: ¶88 and Missionary Evidence before the Commission 1860: 44)). Similarly, as to the knocking down of houses, although a missionary testified that he himself had seen a village '... with only a few houses left ... and the village was so deserted on account of the indigo oppression', the Indigo Commission rejected the testimony as 'not proven', and accepted the planter's word, because 'in one case, particularly, the

Manager of the Bengal Indigo Company, distinctly denied having caused a house to be knocked down'. The Commission did, however, concede that much to their 'fear', demolition still 'occasionally' occurred (RIC 1860: ¶89 and Missionary Evidence before the Commission 1860: 44).

As regards outrages on women—by far the most serious charge brought before the Commission against the planters—both the Commission and the Lieutenant-Government felt that to accuse the planters of outraging women was tantamount to an outrage on decency and justice. The charge was 'disproved', since 'no witness testified to his personal knowledge' (RIC 1860: ¶91). 'The Commission most fully and most justly acquitted all planters', wrote Grant in his Minutes, adding

> it is to be regretted that any charge as gross and unfounded was ever made in any quarter, for it is neither true, nor has it any likelihood of truth in it (Buckland 1901: I, 258).

That women were being raped by planters was known to their friends and relatives in the village and also to some missionaries whom villagers trusted. In *Neel Darpan*, Mitra's story of rape is no figment of his imagination; it may even have been based on the evidence of the Reverend C. Bomwetsch before the Indigo Commission. Giving evidence this missionary had said:

> With regard to outrage on women, the brother of a man whose wife had been taken away from a *ghaut*, came and told me of an incident; the man's story ran thus: Two young women were fetching water at a *ghaut* and were taken away by the servants of a planter. After finding out which was the woman they wanted ... they let the other one go on her way, and the one who was the beauty, was carried to the factory ... (RIC 1860: Missionary Evidence).

The colonial masters, imbued with Victorian puritanism and the ideology of racial superiority, shrank at the very thought of a white man having sexual relations with a native woman, let alone raping her, and worse still of such stories getting publicity. The victims too were not keen on publicity, and the reasons for this hold good in modern India just as much as they did in the nineteenth century. More than the outrage, it was the spreading of the fact of the outrage that was unacceptable to all. In this connection the evidence of the Reverend Lincke is of great importance. The reason he gave in his evidence before the Indigo Commission for having 'no personal knowledge of any of them, though he had heard of several cases of outrage on women' was that

> the people are most careful not to speak of such subjects; for once known that anything had happened to their women, their caste would be gone (RIC 1860: Missionary Evidence).

Reverend Bomwetsch also gave his evidence on the subject and though it was not accepted by the Commission, this was in all likelihood the basis of the story of kidnap and rape we find in *Neel Darpan*, for the circumstances under which the two young women were kidnapped by the planter's peon are identical in the evidence and in *Neel Darpan*. The anonymous correspondent of *The Hindoo Patriot* reporting from the rural areas also had been mentioning cases of crimes against women. On 22 August 1860 this correspondent wrote to his Editor:

> It is a pity that the *Indigo Commission* had no time to come to this district. A hundredth part of the actual oppression has not been reported to the Commission. Even now traces of house-burning, plunder, etc., many persons 'outcasted' for the grossest outrages on their seraglios, can be seen in the Indigo district. ... Here females have been forced to break the clods in the indigo fields in the absence of their husbands; and wives and daughters are carried away and confined in the godown, where they are of course treated with the greatest insolence imaginable, to compel the men to cultivate indigo (Letter IX).

In *Neel Darpan* the planter Wood brags (Act I, Sc. 3):

> You not see, I whipped those mother-fuckers, snatched cattle, locked up the wives? Lock up wives, then bastard people taught lesson.

The Hindoo Patriot mentions another instance of brutality on women by the Hazrapore planter, Oates. In the month of February Ramtanu Adhikari was attacked in his house in the village; Ramtanu fled, but the women could not. The clubmen first demolished the house, looted the furniture, then seized the women; they removed their clothes and forced them to stand naked in the yard. 'I shudder to write the remainder of the tale. Ramtanu has been outcasted. When he complained against Oates in the court of Skinner, he lost no time in dismissing it.' Another family—the Shahs of Lautara—also suffered similarly at the hands of Oates (cf. Bagal 1953). So to say that the charge of crimes against women 'is neither true nor has it any likelihood' was a gross transvesty of truth. Victorian puritanism winked at such aberrations as long as they did not get known. Secrecy had to be observed at all cost, to keep up the facade of British morality. *Neel Darpan's* English translation under Reverend Long's guidance and its circulation in England violated this unwritten rule; it exposed and widely publicized the fact that white planters were forcing native women into sexual intercourse. While addressing the jury in the trial of Reverend Long, the Supreme Court Judge Mordaunt Wells lost his equanimity and observed angrily: 'What public good would it serve to publish these filthy allusions to the prostitution of the native women to the planters?' (Kling 1977: 215). It was not the horrendous crime, but its exposure which upset the judicial temper, and Reverend Long had to pay for this.

THE ROLE OF *ZAMINDARS*

Generally speaking, the big Bengali landholders chose to keep aloof from the peasants' resistance movement. They had a lot to lose by openly antagonizing European officials in the districts, who they knew were friends of the planters. Nor would it have been prudent on their part to forego the enormous income accruing from the *patni* leases they gave them by cancelling these. Even Ramrattan Roy, a terror to the planters, declined to help Srinath Bose, a *beilaqa* peasant of Kushtia, probably his own tenant, when he complained that his land had been forcibly sown by the planter Kenny's men, and that Kenny himself had cut a canal through his good rice land. Roy's excuse was that, in a similar dispute, his own *naib* had been imprisoned, so he could not meddle in this case (RIC 1860: ¶3262). But, for all their caution, there was no love lost between the planters and the *zamindars*. Considerations which weighed with the big *zamindar* Jaykrishna Mukherjee against allowing peasants within his *zamindari* to sell their holdings to European planters, weighed also with other landed gentry. They were loathe to surrender their traditional influence as *zamindars* and were afraid that Europeans would acquire too much (Mukherjee 1975: 235). Jaykrishna Mukherjee's standing advice to those of his tenants, who had been allowed by him to produce indigo for the planters, was

> If servants of the indigo planter oppress you, complain to the *sahib*, if he does not listen, complain to me, and I will settle with the *sahib* (Mukherjee 1975: 235).

But the *zamindars'* condemnation of the Rent Act of 1859 was as virulent as that of the planter-*zamindars*; though the new Rent Act benefitted the landholders much more than it did the peasants, at the time of its passing in 1859 it had raised a storm of protest. Landholders felt that the government was making an attempt to 'level the Native aristocracy to one standard of social equality with the rural classes' (*The Hindoo Patriot*, 5 May 1859). Jaykrishna Mukherjee produced a pamphlet called '*Permanent Settlement Imperilled: Or Act X of 1859 in its True Colour*' (Mukherjee 1975: 259). When their vested interests, they felt, were in danger, there was little difference between the planter-*zamindar* and the local *zamindar*. James Furlong of Nadia, for example, had criticized the Act X, as 'an unwarrantable liberty taken by the Government in interfering with the relations between the landlord and the tenant' (Chowdhury 1964: 181). These men could not possibly have given their support to the peasants' movement, which most of them feared would transform the familiar, docile, obedient peasant to someone aware of his basic rights. But there were a few exceptions. Shri Gopal Pal Choudhury's house in Ranaghat was a meeting place for like-minded *zamindars* to discuss anti-planter views (Kling 1977: 69). His brother Sham Chandra Pal Choudhury was accused by Larmour of being 'the sole cause of witholding rent' during the Rent Disturbances (Montressor 1862). Another *zamindar*, the irrepressible Mahesh Chandra Chatterjee, must be mentioned once again (Kling 1977: 88). By the end of August 1860, when the indigo disturbances had spread to the subdivisions of Magura, Narail and Khulna of Jessore District (Reid 1860c), and indigo was being destroyed on the ground at Faridpur's Paccadanga Concern and the Manager and his agents were being prevented from entering the villages in Katgarrah (Chowdhury 1964: 184), a remarkable demonstration was organized by Mahesh Chandra Chatterjee as the Lieutenant-Governor returned by boat from his tour of Sirajgunj. Thousands of men, women and children collected from distant villages to line both sides of the river as a mark of protest (Buckland 1901: I, 192). What made this demonstration really remarkable was the veiled threat it held out through the extraordinary show of strength of tens of thousands of disciplined, orderly and respectful, yet determined demonstrators demanding change and justice. The significance of the demonstration was not lost on the Lieutenant-Governor. Deeply impressed and uneasy, Grant recorded his impressions in his Minutes (1860):

It would be folly to suppose that such a display on the part of tens of thousands of people has no meaning. The organisation and capacity this remarkable demonstration over so large an extent of the country are subjects worthy of consideration.

The significance of the reported demonstration was not lost on the Governor-General of India Lord Canning, either. Alarmed, he wrote to Wood, the Secretary of State, India, that the demonstration had 'caused me more anxiety than I have had since the days of Delhi' (Buckland 1901: I, 192). He continued:

A people who can do this, and do it soberly and intelligently, may be weak and unresistful individually, but as a mass they can not be dealt with too carefully. ... From that day I felt that a shot fired in anger or fear by one foolish planter, might put every factory in Lower Bengal in flames (Canning 1860 in Kling 1977: 169).

A startled government sent two gun-boats to the rivers of Nadia and Jessore, and despatched military police and native infantry to the two indigo growing districts (Buckland 1901: I, 193). An overwrought Governor-General asked his Lieutenant-Governor:

to what places beside Nadia it would be advisable to send troops ... and to what extent to have the Native troops in Bengal reduced as compared with the number which we had in April 1857. The European troops were to continue at every station which now has them (Canning 1860).

Canning was 'glad' that 'there was a good force in hand against the rioters', in case there was trouble during the October sowing. The notification, which had been promised to the peasants to assure them that they would not be coerced to sow indigo after the expiry of Act XI, 1860, was not considered 'to be prudent to be made publicly known before the troops show[ed] themselves' (Canning 1860). Before it was issued, the Notification was suitably amended by the Governor-General to '... disabuse the minds of the rural population of an erroneous impression that the Government was opposed to the cultivation of indigo' (Buckland 1901: I, 192), and also 'to warn the *ryots* against intimidating anyone who wished to sow indigo' (Kling 1977: 169).

The disturbances, however, continued and had spread from Nadia to Faridpore, the stronghold of the Fara'izis, who had been active in Sirajganj in 1859, directing the peasants in the *zamindar*-planter conflict. Towards the end of October 1860 they struck again. United under the leadership of the Wahabi Rafiq Mandal (Tara Chand 1967: II, 534), the peasants refused to cultivate indigo and assaulted the factory workers and servants (Buckland 1901: I, 193). But by this time the dreadful Act XI had been withdrawn, and a proclamation declared assuring the peasants that 'they would never be compelled to sow indigo'. The quasi-serfdom of the *beilaqa* peasants came to an end, and it was for those who were still in bondage to take up the struggle in the *ilaqa* areas of the planters.

The summary law, known as the Indigo Coercion Act, had served its purpose, for the greater portion of the indigo sown in spring had not been lost. A report from the magistrate of Nadia, quoted in Chowdhury (1964: I, 184), stated: 'It is tolerably clear that the *ryots* had yielded'. It was noted with satisfaction that 'Less than one-fifth of the produce of the Bengal Presidency, or about one-seventh of the produce of all India has been in peril' (Canning 1861). Grant also stated that 'It will be an excessive calculation to say that twenty thousand *maunds* are in jeopardy, owing to the inadequate price offered to *raiyats* in Bengal' (Minutes in Buckland 1901: I, 257). Out of '60,000 men employed in the cultivation of indigo in one season or another' (Canning 1861), many clearly had submitted to the planters in carrying on cultivation. They had been compelled either by the pressure of circumstances or by the ferocity of the law and fear of oppression. But '*ryots* brought to trial did not give in' (RIC 1860: ¶2819); their anger had not subsided. If anything, as Herschel observed, 'the feeling that existed formerly, though not aggravated, is ten times more determined' (RIC 1860: ¶2819). Perhaps surprised by their own spirit of unity, which they themselves had never known before, the peasants took it as a gift of God; as Reverend Lincke (RIC 1860: ¶918) said, 'it is from God and the time for release had come'. Sir Bartle Frere stated rather prophetically,

> The question was far graver than its merely economical aspect. Regarding it as only one symptom of a general awakening to the condition of serfdom under which they have long existed, the crisis does not seem to be yet passed (Frere 1861).

By the end of 1860, when the indigo resistance movement took a new turn, and became what is known as the 'No-Rent Campaign', *Neel Darpan* had already been published and widely acclaimed by the public. As such, any detailed analysis of the rent movement has no place in this study.

VI

THE NEEL DARPAN *CASE AND REVEREND JAMES LONG*

The most outstanding missionary in Bengal during this period was James Long of the Church Missionary Society. Entering Bengal in 1840, he had mastered Bengali and earned the reputation of being a distinguished scholar in that language. Today copies of all original publications in Bengali which he had procured are in the India House Library and his collection of all Bengali translations of Sanskrit texts are at Oxford. Long's most active and abiding interest was in the vernacular press, which he felt was 'an exponent of the views and feelings of the masses'; with careful selections from various newspapers he compiled three volumes in Bengali. He believed that vernacular literature, newspapers, pamphlets and all other items 'already in circulation among the native population should be translated (into English), for the information of those to whom it was of importance to understand native feelings' (Long 1856: 85).

This became the mission of his life. True to his own convictions, 'when new vernacular books of special interest appear[ed]', he was 'in the habit of sending them as indices of Native mind to Missionaries, Educationists and Government Officials ... in order that such persons may observe the current of popular feeling'. With obsessive concern he noticed that 'a small number of Europeans' lived 'in the midst of a vast body of Natives' and feared that 'a volcano may be forming beneath their feet, and dark clouds may be gathering on the horizon of India'. His advice was that 'for the future security of Europeans in India ... all classes of

Europeans should watch the barometer of the Native mind'. In other words, keeping the authorities informed of what was going on beneath the surface was one's primary duty as an European, and since 'few Europeans know anything of the Native language', they must be given translations to understand native feelings. 'Had translations been made from the Native Press of the Agra Presidency, indicating the state of feeling towards Government a year before the Mutiny ... Europeans might not have been taken so unawares ...'

These were the thoughts which made it imperative for Long to show the play *Neel Darpan* to Seton Karr, then Secretary to the Government of Bengal. Reverend Long regarded this play 'as an expression of Native mind', for the general public he felt had come to believe that 'the effect of the indigo system on Native families had been as ruinous as the drama represented'. Whether the facts represented in it 'are entirely true or partially so', they ought not to be trifled with. Seton Karr read the play in Bengali, and like Long saw the publication as informative for the white population of 'potentially dangerous Indian opinion'. The Lieutenant-Governor also wished to read the play, because he 'thought it likely that it would show what the real popular feeling was on the subject better than anything else'. So under Long's guidance the play was translated into English anonymously by Michael Madhusudan Dutt. But instead of publishing a few copies which could be given 'to friends in official and private positions', as Grant had instructed, before leaving for Parishnath on tour, Seton Karr got five-hundred copies published, due to some error. One hundred of these were sent, under government's frank, to eminent persons both in India and in England.

Furious planters denounced the play as 'infamously obscene and grossly libellous', and effectively checked its wider circulation; hundreds of copies remained undistributed in the Bengal Office. On receiving a copy from the editor of *The Englishman*, the Secretary of the Landholders' and Commercial Association of British India, W.F. Fergusson wrote to the Government of Bengal on 25 May, 1861, asking if this 'foul and malicious libel on indigo planters tending to sedition and breaches of the peace' was distributed officially by the government. In his reply of 3 May, 1861,

Grant regretted that 'a misunderstanding [had] been caused', but he defended the translation, observing that

> ...it is Vernacular Plays, not English translations that may be dangerous, nor can the knowledge of respectable officials or other European gentlemen of the existence of such indications of popular feelings as this be anything but a security against actual sedition and breaches of the peace.

TRIAL OF JAMES LONG

The planters were not in a mood to accept 'regrets'; they wanted their pound of flesh. When legal proceedings were instituted on the charge of libel and E.H. Manuel—the printer of the English translation of *Neel Darpan*, whose name was the only one on the title page—was indicted, Reverend James Long accepted responsibility for having the translation published. The charge against Manuel was then dropped with a fine of one rupee, and Reverend Long was indicted; on 19 July 1861 he was tried for libel before the Supreme Court. Strangely, the prosecution did not name the planters as the aggrieved, but used instead the name of the editor of *The Englishman*. As the *Calcutta Christian Observer* stated,

> They must have thought, that as regards the Planters, there was very little ground for a prosecution, or else this course of putting forward *The Englishman* never would have been adopted.

Long made a statement in 'self-defence' on 20 June on whose appearance,

> ... most moderate men in Calcutta, including not a few of the merchants and even some members of the Landowners' Association, wished the prosecution to be discontinued. But the counsels of the 'more violent' prevailed and on 19 July the Supreme Court beheld its first State trial (*Calcutta Christian Observer*, August 1861).

Prior to his trial, distinguished citizens of Calcutta presented an address to Reverend Long; another address, signed by three-thousand Indians, was sent to him subsequently (cf. Appendix II).

It was one of those cloudy days of Calcutta in the month of July. The courtroom was crowded, the whole of Calcutta was there: European and Indian officials, missionaries, merchants, bankers, indigo planters and the 'stout figures of influential natives' (Kling 1977: 203). Sir Mordaunt Wells was the presiding judge, well known for his statement from the Bench on 24 August 1859 that the 'entire nation of Bengalis were forgers and perjurers'. Also known was his request that an Englishman, convicted of beating to death his old Indian servant, should be treated with leniency (Kling 1977: 210). There were fifteen jury men, of whom all except L.A. D'Souza and Manickjee Rustomjee were English. The opening remarks of the prosecutor were significant; he discovered some errors in the translation, which he felt proved that it had not been done 'by a Native'. The charges were: first, that the Preface to the play slandered the editors of *The Englishman* and the *Bengal Hurkaru*; second, that the text of the play slandered the planters as a group as well as English womanhood. Remarking that the pamphlet was written with a view to creating discord among the races, he played upon the deepest fears the British had since the days of the Mutiny of 1857. None of the members of the jury could have forgotten the Panic Sunday of 14 July 1857, when, though 'no one had laid a finger on Calcutta, it was plain terror mixed with shame, and it became racialism of a kind the city had not known before' (Moorhouse 1984: 78). 'Had we not seen by what a tender thread we hang? Have not the late Mutineers taught us how unsafe is our position?' argued Peterson. The sentiment expressed in this statement was no different basically from Long's own utterances regarding the insecurity of Europeans living in the midst of 'a vast body of Natives'. But the argument of the adroit prosecutor was given a sinister twist when he observed:

> If that reckless slander had been promulgated among the war-like tribes of the North-West it must have inevitably led to the extermination of the Europeans.

With a cold shudder running down their spines, the jury men could not have agreed more readily.

Under the law of libel then prevalent in India, Long was not permitted to argue his own case, and unfortunately the defence lawyers were not up to the mark. Eglington showed 'a marked want of ... nerve and presence of mind necessary in a jury trial' and Newmarch was a 'worn out old man, said to be very agreeable up to twelve o'clock but apt to be muddled by intemperance when most wanted' (Kling 1977: 203). But when the judge had taken upon himself the role of a prosecutor-cum-attorney, the defence could not have convinced the Bench of their client's innocence. Sir Mordaunt Wells told the jury that he would not pronounce whether the work was a libel or not, but state the law and leave the matter entirely to them. Yet, he went on, '... in a manner rarely witnessed on the Bench, to denounce both Mr Long and the work', which he called 'a foul and disgusting libel'. He read out the passage 'which mentions a European lady' (cf. Act IV, Sc. 3): 'I will read it', he said, 'it is set out in page sixty-six of the pamphlet'. Then he went on to explain,

That is the wife of Mr Wood, and through her influence the bearer got the *jemadari*-ship, and the Magistrate was the only person with whom she would dance.... Now, whether this would lead fairly to the inference that Mrs Wood would improperly influence the Magistrate, is a question which the jury have already decided. ... In my opinion it is incapable of bearing any other interpretations. ... It appears to me that it is doing as great an injustice to the Indigo planters to say that they used an improper influence over the Magistrates through their wives, as it would be to say that the Judges of this Court had, while suits were pending, received such a letter from one of the litigant parties. Then the question is whether there is anything in the libel to show any imputation ... had been cast to bring into contempt the administration of Justice in the *mofussil* (*Calcutta Christian Observer*, pp. 90–1).

The judge read out to the jury various passages from the play, and as the prosecutor Peterson had done, '... pursued with a righteous indignation the libeller who had dared cast the deepest stain upon the fair fame of his country women, the English womanhood ...'. It was, he continued, impossible to speak of these passages in the play '... otherwise than as filthy insinuations against a society of helpless ladies, who under the mask of a general type, were cruelly stabbed in the back.' He also read out the Preface, and asked if 'respectable gentlemen who conduct newspapers in Calcutta ... would sacrifice the welfare of Society to the promotion for a corrupt purpose of private interests? If it is meant that would it not be a libel?' If the jury believed, continued the judge, that 'the defendant had been actuated by a feeling of animosity towards the planters of Lower Bengal, with a view of degrading, injuring and bringing this class into contempt and ridicule ... the verdict must be guilty.' The jury obliged the court. Long's attorneys requested for 'arrest of judgement', and on 24 July the full Bench presided over by the Chief Justice, Sir Barnes Peacock, heard the case. Long was allowed to make a statement before the full Bench. His argument, full of emotion, sought to justify the publication of *Neel Darpan*: that the *raiyat* was a dumb animal, who did not know the rulers' language, that at the time of first publication matters on the indigo controversy were assuming such a threatening aspect, that it was important that men of influence should know that 'the wound was not a surface one, but required deep probe'; all this, however, had no effect on the decision of the august Bench. The Chief Justice interrupted Long midway, explaining that 'it was not the length of the matter he was now reading, but its substance they objected to, as irrelevant' (cf. Appendix for Long's entire statement). Peacock upheld Wells' decision, and Reverend Long was sentenced to a fine of Rs 1000 and one month in the common jail. The fine was paid for Long by Kali Prasanna Sinha, a patron of literature.

The sentence shocked the public of Calcutta, and even those newspapers which did not support missionary activity, wrote angry editorials. *The Hindoo Patriot* (8 August 1861) compared 'the judicial prosecution of the Reverend Long to the judicial murder of Nundcoomar'. In a long editorial entitled 'The Revolutionary Trial', it wrote,

> In Nundcoomar's trial the most extraordinary thing ... was the demeanour of the Chief Justice ... who was determined at all hazards to relieve forever his friend and head of the Government of the one man who brought that Government all but to a dead lock. ... Throughout Mr Long's trial the presiding judge discovered a kindred spirit. He did not ... exact an investigation into the real point on which a case of libel, or in fact a trial for almost any offence at law hinges, but allowed and ... encouraged a useless and impertinent discussion ... and when a blustering and bullying charge induced a verdict of 'guilty', he pronounced a

sentence the savage severity of which took even us—who flattered ourselves with possessing in our mind a full measure of Mr Justice Wells—quite by surprise.

Regarding the jury, the paper commented with sarcasm that

the credit of the blunder of the verdict does not belong to the jury. The conduct of the presiding judge and the extraordinary summing up virtually dictated a verdict to men who are never guilty of opinions of their own (*The Hindoo Patriot*, 8 August 1861).

Barring *The London Times* which merely reported the trial, the British press also strongly criticized the judgement. *The Daily News* wrote: 'It was as if the French clergy has prosecuted Molière, or the Yorkshire school masters the author of *Nicholas Nickleby* or the Southern Planters the author of *Uncle Tom's Cabin.*' *The London Review* urged 'the dismissal of the judge whose sentence on the defendant shows a spirit of partisanship which is never witnessed on the Bench in England and can not be tolerated in her dependencies'. Among the officials, Bartle Frère was most fortright. On 9 August 1861 he wrote to Wood, the Secretary of State, saying that 'the sight of English judges behaving as (Peacock) and (Wells) have done, throws everything in the shade' (Kling 1977: 209). Lord Canning also called Wells' behaviour 'indecent', and Peacock's interruption of Long's defence statement 'a discreditable performance' (Canning 1861: July 31). Yet Canning rejected the petition signed by over 20,000 people, for the recall of Wells for his 'repeated and indiscreet exhibition of strong political bias and race prejudices....' (Moorhouse 1984: 73). The rejection was accompanied by the usual expression of regrets, and with the hope that with 'time and reflection' the emotions of the Bengalis would subside.

After his month in jail Reverend Long left for England. But the wrath of the planters had not subsided. They were planning to collect evidence against Seton Karr for his involvement in the publication and circulation of the translation of *Neel Darpan*, and his conviction was almost certain, as the Advocate-General had declined to appear in case the matter came up before the Court. But the Governor-General's resolution condemning Seton Karr was published on the day set by the Landholders' and Commercial Association to take their final decision. Thus, Seton Karr was saved from imprisonment, but he resigned from his secretaryship in the Government of Bengal; he was subsequently appointed a judge of the High Court and still later became Secretary to the Government of India in the Foreign Department. J.P. Grant also was not spared the planters' anger; he too was censored by Canning and was more severely punished. An assistant in the Luckipara indigo factory of the Mirpur concern brought a suit for libel against Grant for a report by Lushington, at the time Commissioner of Nadia. The report mentioned that the planter McArthur as well as one Driver, manager of the concern, probably knew and assented to the attack by their servants on one of the followers of the *zamindar* Hurnath Roy, which culminated in this man's death. The planters claimed not to have been there when the murder took place, and as the report was an official publication Grant was named in the libel suit. Peacock presided, found Grant guilty and fined him one rupee. Grant left India in April 1862.

VII

NEEL DARPAN *AS A SOCIAL MIRROR*

THE STORY

Neel Darpan tells us the story of Swarpur, a village by the Ganges, somewhere in the district of Nadia. It is 'a village of cultivators', as the milkman puts it, and in his eyes it is not much of a place for 'a city gentleman' to give his daughter to in marriage. But to the small landowner, Golak Chandra Basu, it is his 'golden Swarpur', where his ancestors lived for seven generations and left enough land for the family to live off in moderate comfort, without having to look for employment in the city. Before the planters, I.I. Wood and P.P. Rogue had entered the village as *patnidars*, Golak Chandra Basu's barns used to burst with rice, enough for the family and for whosoever dropped in, all the year through. His lands also yielded 'lentils, molasses, mustard seeds for oil and fresh vegetables', and the ponds had abundant fish. The annual Durga Puja festival used to be celebrated with great pomp and rejoicing; money used to be given away in charity, to Brahmins and the poor; the houseful of guests used to be feasted and the villagers regaled with the performances of the travelling Vaishnavas and the *jatra* parties.

Gone were these days ever since the planters took lease of the village. Not only was there economic ruin, even socially the planters spelt disaster. No village woman was safe, once planter Rogue had cast his eyes on her while riding by. Pretty, young, pregnant Kshetramani, the only child of Rebati and Sadhu

Charan, catches Rogues's eyes; she is kidnapped and raped, and dies in great agony.

Before long Golak Chandra Basu loses all his free-hold land and his entire ancestral property. He is not even paid for the indigo already delivered to the planters, but he has to face a fresh demand to cultivate the plant on the few remaining *bighas* of his rice-land. His son, Nabinmadhab, a man with modern education, takes the matter to court, since he is sure of getting justice done. The new magistrate is, however, a friend of the planters, and has made up his mind even before hearing the arguments. Using the newly passed Act XI of 1860, he sends the old father to jail on the planters' false charges that he had been inciting the peasants not to cultivate indigo. Unable in his old age to bear this humiliation, Golak Chandra Basu first goes on a fast and then hangs himself inside the prison.

When, on his return to the village, Nabinmadhab hears that the planters are going to sow indigo all around his private pond, he goes to Wood and begs him to wait a few more days, till his father's funeral is over. Wood not only refuses, but thrusting his boots towards him, he offers them as gifts for the funeral. Furious at this obscenity, Nabinmadhab Basu kicks him hard and throws him down. The planter picks himself up, and breaks Nabinmadhab's skull with a blow of his stick. The unconscious Nabinmadhab is picked up and carried home by Torap, a loyal tenant; there he dies. The grief stricken old woman, Sabitri, who has so recently become a widow, loses her mind when she hears of the sudden death of her son. In a fit of madness she strangles her younger daughter-in-law, Saralata, whom she was so fond of and who had been looking after her since her bereavement. Then, regaining consciousness for a moment and realizing what she has done, she herself dies of shock.

This bald narration does not convey half of what *Neel Darpan* revealed: the cruelty, the arrogance of the colonizers, the suffering and anger of the colonized; the poverty of the illiterate, but also their capacity for laughter, their raw, uninhibited articulation of their perception of the reality. Dinabandhu Mitra did not seem to care whether these cultural expressions of the indigenous peasantry hurt the sensibilities of the nineteenth century gentry of Bengal.

THE QUESTION OF AUTHENTICITY

While growing up in Nadia's Chanberia village, Dinabandhu Mitra had most probably already learned that the entire indigo industry was based on forced labour, and that it rested wholly on a foundation of coercion and intimidation. As a boy, he must have heard about those gruesome crimes committed by the early planters in the countryside, for which in 1810 the Governor-General in Council, Lord Minto, had deprived four of them of licences to reside in the interior of Bengal. The story of Golak Chandra Basu is in all likelihood based, as the *Bharat Samskaran* (7 November 1873) mentioned in Mitra's obituary, on a similar incident in some Mitra family in Nadia's Guateli village. In later life, as an inspector of the postal department, Mitra was—as he mentions in his Introduction—really a 'traveller'. Especially after his transfer from Orissa to the Nadia Division—the very heart of the indigo-growing area—he could be more involved in the observation at close quarters of the methods of oppression, of the behaviour of the magistrates, of the reaction of the villagers, of the general demoralization and corruption from the court officials right down to the planters and their employees. Conversant with the local dialect, it could not have been difficult for him to collect authentic information. This information together with the evidence presented before the Indigo Commission provided the basis for *Neel Darpan*.

THE VILLAGE

Swarpur is a replica of the larger society to which it belongs. It has the same complexities of status, caste, religion, profession and economic power. There are as many castes and sub-castes as there are professions—village doctors, traders, cultivators, milkmen, oilmen, fishermen, clubmen, domestic servants and prostitutes. Then there are the modern, educated professionals—teachers, physicians, lawyers, court officers and the hundreds of employees of the planters. There are the planters themselves—arrogant white men using a kind of pidgin Bengali generously strewn with obscenities, daily offending the sensibilities of the people among whom they live. As a contrast to them are the missionaries, who befriend the oppressed. There is a distinct division of labour. Even among the Brahmins, the highest caste, some are priests, some teach in colleges, some in the village school where only Sanskrit is taught; but all are aware of their superior caste status. When offered hospitality for example by the Basu family—also high-caste, but Kayastha—they accept it with condescension, as a mark of favour towards their hosts. Sabitri, the Kayastha widow, is content to break her ritual fast by eating the leavings off a Brahmin priest's plate; but because of her own relatively higher caste, it does not even occur to her to allow Rebati and her daughter of the caste of cultivators to fetch their drinking water from their pond. Even language depends on one's caste—the words, the pronounciation are caste markers; when Sadhu Charan, a cultivator, uses a term not normally used by men of his caste, he is grossly abused and insulted because he has transgressed into the territory of the village elite. A Muslim is inferior to a Hindu, so that Torap may not stand on his Hindu friend's shoulders to peep into the adjacent cell of a planter's lock-up.

Gender relations in Swarpur also conform in every detail to existing norms. Shy and self-effacing, with their *saris* covering their heads and drawn over their faces, women stay inside their homes, attending to the needs of their elders and the menfolk. The mother-in-law's decision is absolute, and the literate young wife Saralata may not reply to her husband's letters, as his mother has forbidden her to. Young Kshetramani is not allowed to wear the little piece of jewellery on the parting of her hair because her husband's mother and elder brother think that such ornaments are fit only for the very rich or for prostitutes. The planter's wife, who goes riding and is free to move around and talk to men, is a contrast to show that women can have a life of their own. Younger men in the village are beginning to appreciate a woman's worth, but there is a long way to go before a woman recognizes her own value, instead of voluntarily merging her identity with that of her husband, even up to the point of wanting to be a *sati*. But Mitra has not painted it all crudely in black and white. The husband-wife relationship is a tender one. The old maidservant Aduri misses her dead husband and describes how loving and caring he was. Bindumadhab's love and longing for his wife, Saralata, come through in his letter, as well as in many little bits of conversation. The relationship between women too is warm. The two sisters-in-law are fond friends; even Podi, the prostitute, rejected by the village at large, is shown a little affection by Kshetramani.

Swarpur is also not isolated from the mainstream of Bengali life. A village girl is given in marriage to a boy in some distant village; a literate city girl comes to the village as a bride; traders bring their wares to the weekly market, friends from neighbouring villages are invited to ceremonies and special feasts and performers come to entertain village audiences. The impact of modern, anglicized education is beginning to be felt; the works of Shakespeare are in demand, children's schools are being planned by the villagers. People are getting aware of certain social and political reforms; even illiterate village women have heard of the great contemporary social reformer Vidyasagar, and the literate read out his writings in favour of widow remarriage; young urban lawyers, visiting the village to report back to newspapers on the oppression of planters, leave behind ideas of resistance.

But it is also the picture of a demoralized society, which keeps silent when a brother forces his sister into the planter's bed, only for his own advancement. The prostitute Podi has no choice but to provide the planter with young women; it is the need for sheer survival which overcomes her genuine distaste for her job. Servile employees, even after being kicked and humiliated, go on flattering the planters. They help the planters in all their illegal and ruthless activities, from collecting false evidence against honest men, to starting false cases against them which take them to jail. It is they who cook up accounts, so that the peasants remain forever in debt, their liabilities passing on to their sons and grandsons. As planters' agents, they take forcible possession of the peasants' lands, drag them to the planters' office, compel them to take advances and, if they refuse, beat them and lock them up, sometimes leaving them there to die. They accompany planters to plunder villages, burn down huts, destroy ripening crops.

The Aim of the Play and The Mirror of a Revolt

The purpose of the play was much more than a mere narration of tyranny and misery. *Neel Darpan* was the first drama in Bengali literature, which sought explicitly to awaken the public, through the symbol of a colonized village, to the painful reality that the entire country belonged to a small but dominant group of foreigners, claiming the right to exclusive possession of power—the power to dictate to the subordinate native community its obligations and duties.

Gopinath Das, the planter's manager, compares his own lot to that of his masters, in the eyes of the unjust law: there is one law for his master, the white man, and another for the people of the country; the white man, the real instigator of the heinous crimes, gets off, at most paying a fine, but his employees go to jail. Within Gopinath, a feeling of guilt has been growing for betraying the family he secretly admires for its courage to stand up against the white man and fight for the oppressed. He feels embarrassed to face the village, where he is looked down upon as the planter's man. He picks up courage to explain to the planter why the style of functioning of the village moneylender is better than the planter's: between the two extortionists, he has cast his lot with the indigenous one, not because the moneylender is good and Gopi approves of him, but because that man springs from the soil and thus relates to the peasant.

With infinite sadness old Golak Chandra Basu tells the court, 'The *sahebs* own this country now ... can we afford to go against them?' (Act IV, Sc. 1) But his son, Nabinmadhab, is not going to yield to the planter's unreasonable demands: first pay me, then talk about sowing indigo, is his ultimatum to the planter, and he thinks of going to court to seek redress. This young man is an example to Swarpur's peasants, their source of strength. Illiterate Ray Charan threatens to go to court after the planter's overseer marks out his best land. The peasant who was picked up from his field and taken to the planter's office on a false charge of breach of contract tells the guard, '... I'm coming, I'm not afraid. I'd rather rot in jail than sow indigo for that bastard' (Act II, Sc. 3). Defiance is in the air, and *Neel Darpan* mirrors it. Grinding his teeth, Torap tells his friends

> ... if I get hold of that swine once ... I shall give him such a punch, that his jaw will go flying to the skies! (Act II, Sc. 1)

Even the village milkman is waiting for the day when the planter's bungalow becomes his graveyard. Gentle Rebati, a peasant's wife, is wild with rage when she hears the planter would offer them money if they sent him their daughter:

> ... don't poor people have their religion! Or their religion has no value, because they're poor! ... Damn money—can one sell one's religion? Can it ever be priced? (Act I, Sc. 4).

Songs are sung everywhere, full of venom ridiculing the planter. The village is seething with anger; those once dumb have found a voice, and Mitra echoes it.

In a previous analysis of *Neel Darpan* (Guha 1974–5) it has been suggested that the play is a 'loyalist' piece of work, 'glorified as a manifesto of bourgeois radicalism', its real purpose being to exonerate the colonialists, since its author had 'no intention of harming the prospect of his professional advancement'. One may then ask why the colonial government failed to appreciate the loyal panegyric; why did it instead send Reverend Long to jail and deprive the editor of *Vividhartha Samgraha* of his stipend for his analysis of the drama. Guha himself wonders why Shivnath Shastri, a well-known Bengali scholar and critic of the time, should have recorded that the impact of this play was like that of 'a violent earthquake ... [which] shook Bengal from one end to another'. If *Neel Darpan* were indeed a display of loyalty, why then did the well-known writer Bankim Chandra Chatterjee pay a tribute to its publication as 'an act of exceptional courage'? Dinabandhu Mitra's ambitions were no doubt limited, and did not extend to projecting the image of a mass uprising of peasants wanting to do away with the entire colonial system; however exhilarating this image might have been, it would not have corresponded to reality. The few words in Sanskrit, introducing the play, made it abundantly clear why and for whom the play was written: it was to comfort the peasants so grossly oppressed by the indigo peasants. Guha finds that the peasant Torap is 'like the "baboo" who created him', and 'has nothing in common with a peasant up in arms'. Reports by Sisir Kumar Ghose, the most reliable correspondent of *The Hindoo Patriot,* show, however, that the general mood of the peasants on the eve of the clashes was relatively passive. 'The pressure [of Act XI of 1860] was too much for the *ryots* to withstand, and as a consequence there were clear signs of weakening of the resistance movement....' (Chowdhury 1964: I, 184). Guha also accuses Mitra of having a 'communal, parochial....' outlook; he does not give any instances in which this outlook manifests itself, but what he means is perhaps, that Mitra portrays the communal and parochial outlook of village society—that the mirror reflects the reality. While telling us the story of the destruction of a small landowning family and the ruin of a tenant's happiness at the hands of the planter, Mitra conveys much more. Throughout the play he conveys his own sympathetic comprehension of the world of the neglected, the illiterate, the exploited from their own angle of vision; entering the recesses of their minds, he has told the story from below.

The Staging of *Neel Darpan* and the Reactions

The history of the play *Neel Darpan* is closely linked with that of the growth and development of public theatres in Bengal. The first rehearsals of the play took place in the Baghbazar Amateur Theatre, which was a room in the house of Bhuban Mohan Neogi, on the banks of the Ganga (Mukherjee 1982: 31). On 19 November 1872 an advertisement appeared in the *Sulabh Samachar* which mentioned that the Calcutta National Theatrical Society would stage the play on Saturday, 7 December. This marked the opening of the first Bengali public theatre, at No. 33 Upper Chitpore Road, in the house of Madhusudan Sanyal. The performance took place by gaslight and ended at midnight; the audience were seated on chairs, benches and on the steps of the courtyard staircase. The tickets were priced at one rupee for the first class (chairs), at 8 *annas* for the second (benches) and at four *annas* for the third class (steps). The parts of Planter Wood, Golak Chandra Basu, Sabitri and a *raiyat* were all played by the well-known actor Ardhendu Sekhar Mustafi. For the first time the elite of Calcutta saw on stage the rude sons of the soil, instead of highly dressed up kings and queens, Gods and Goddesses; instead of high flown eloquence, they heard a peasant dialect. All along theatre lovers in Calcutta had had to depend entirely on the munificence of rich patrons; to be able to produce a play on their own was a distant dream. With *Neel Darpan* this dream became a reality, for the production—needing neither expensive apparel nor impressive backdrops—cost next to nothing. Villagers appeared in their ragged, everyday wear, and the Basu family was no different from any other middle class family in Bengal; all that had to be procured were a couple of trousers and boots for the planters.

The play was, as is to be expected, received differently in different circles. The Bengal press published rave reviews. *The National Paper* commented (11 December 1872) that 'it was an event of national importance'. *Sulabh Samachar* (10 December 1872) wanted 'more such plays for the welfare of the people'. After praising the play and the performance, the *Amrita Bazar Patrika* (12 December 1872) suggested that 'it should be staged outside Calcutta—like Krishnanagar, Berhampore, Jessore where the contents would be more relevant and valuable'. The *Bharat Samskarak* described Mitra as one of the greatest writers of tragedy. There is a story, which could be purely apocryphal, that while witnessing the play, the famous social reformer Ishwar Chandra Vidyasagar was so deeply stirred and carried away by its stark realism, that he took off one of his slippers and hurled it at the actor playing the part of Rob Rogue. Another great contemporary of Mitra, the author Bankim Chandra Chattopadhyay, commented not only on the author's 'deep humanism and his extraordinary knowledge of village society', but also on his forceful writing. Shivnath Shastri, a famous name in Bengali literature, expressed what most of Bengal felt about the play in almost ecstatic language:

> The terrific excitement the play had aroused in every part of Bengal is unforgettable. Old and young, men and women, all of us were engrossed and were swept off our feet. It was the only topic for days and it was staged everywhere in every locality. From one end to the other, the whole of Bengal shook as if there was a violent earthquake.

Not surprisingly then, *The Englishman*, the planters' paper, was critical (20 December 1872) for political reasons. It wrote:

> Considering that the Rev. Mr Long was sentenced to one month's imprisonment for translating the play which was pronounced by the High Court as a libel on the Europeans, it seems strange that the Government should allow its representation in Calcutta, unless it has gone through the hands of some competent censor and the libellous parts excised.

It seems a performance, while in progress in Lucknow, had to be abandoned as the British members of the audience rushed to the stage in anger, when Torap, in the rape case, forced his way into the planter's apartment and beat him up. That the play excited the British is clear.

Appendices

APPENDIX I

Selections from papers on Indigo Cultivation in Lower Bengal

1. *The ryot as a free agent*

It would appear from the report of the Indigo Planters Association Meeting (of June 18, 1859), that some evil-disposed magistrates, being moved and seduced by the instigation of the devil, have been daring to talk to the *ryots* of rights. Well may Mr Theobald be indignant, and talk of a petition to Parliament. The idea of a *ryot's* rights!! Is it not laid down in Magna Carta, that a black man has no rights? Well may Britishers fear that we are about to lose India from the incompetency of our rulers, when magistrates begin to talk about a nigger's rights. As Mr Theobald indignantly observes, 'for Magistrates to encourage what is at once contrary to law and morality, and to the fair interests of capital, is intolerable....'

Now we all know, at least we have all been told, time after time, that the Secretary to the Indigo planters has a monopoly of legal knowledge; he has taken every opportunity of impressing upon the public, that he is the only man in India who knows anything whatever about the matter.... We shall not therefore of course endeavour to dispute a question of law with so distinguished a jurist. We only proceed to *ad questionem facti;* is it Government treating the planters thus and the planters treating the *ryots* as they do now. There is no more a contract in the one case than in the other.

—*The Indian Field*, 26 November 1859

2. *Description of giving advance to a ryot*

Very generally the advance-money is paid by the native *Gomashta*, but in some cases to meet conscience and public opinion the task is undertaken by the superintendent himself. The *ryots* crowd in the open compound before the verandah where the *saheb* sits in a chair, surrounded by his native subordinates, who squat on the floor with the account books, and with purses containing money. The *Mohurir* of the factory then calls out the name of a *ryot* from the list before him and then sings aloud to the effect:

—'You, Nazir Mahmood of Neelcooteepoor, last year you cultivated two *beegahs* of land with indigo, but as by the report of the *ameen*, it appears that you have purchased additional cattle, you are to cultivate one *beegah* more this season, you are booked for three *beegahs* from this year. But the factory accounts show a balance of four *rupees* against your name, so deducting that sum from six *rupees*, which is the amount to which you are entitled this season for three *beegahs*, there are left two *rupees*, from which again deduct two *annas* on account of a stamp paper for your *kubooleut*, twelve *annas* for the price of indigo seed to be supplied to you and six *annas* for the repair of roads ... in all making a total of one *rupee* and four *annas*, the sum which you are now to receive for the cultivation of three *beegahs* amounts to twelve *annas*. Now come forward and take it, the *Saheb* will pay you with his own hands.'

No sooner is this speech ended, than 'Dohye Saheb' Nazir Mahmood screeches out 'Do not burden me with an additional *beegah*, for I will not be able to cultivate so many as three *beegahs*. I came here with the purpose of soliciting you to relieve me of one of the two *beegahs* which I have hitherto sown with indigo. For God's sake, take mercy on me, *Saheb*; Allah has taken away from me only two months ago my eldest son, who was the prop of my old age ... he it was who ploughed your indigo lands and it was through his exertions that my whole family got a mouthful of paddy everyday to support our strength. ... The new bullock reported by the *ameen* was purchased by my son before his death, but has been mortgaged by me to defray the expense of his sickness and funeral ...'

'*Neel hoga*' is the brief answer of the *saheb* ... the *tageedir* and the *chuprasee* bound towards him and after a shower of blows and slaps, pull him forward.

—*The Hindoo Patriot*, 31 March 1860

APPENDIX II

Address by 'the principal Natives presented to Mr Long' before the trial (CCO 1860: 88–9):

TO THE REVEREND J. LONG

SIR, —We, the undersigned, have persued with attention the Statement, which you have lately published, explanatory of your connection with the *Nil Darpan*, a work of fiction, illustrative of the feelings of the people of Bengal, on the subject of Indigo Planting, as carried out in this part of the country.

The part which you have for years together taken in the advancement of Vernacular Literature and in the dissemination of the views and feelings of the Natives on topics of administration and social improvement, *as reflected through the medium of the Vernacular press*, has justly entitled you to the gratitude of all classes of the native community, notwithstanding the difference of religious sentiment between you and them; and we believe the cause of good government has not been a little furthered by your industrious application in bringing those sentiments and feelings to the knowledge of the governing Authorities, and the local European Public.

Constituted as the British Government is, it is needless for us to dwell on the importance of consulting in matters of legislation and administration, native opinion and native feelings expressed in whatever form and through what medium soever, but we beg leave to state that we fully endorse your opinion that '*peace founded on the contentment of the native population is essential to the welfare of India, and that it is folly to shut our eyes to the warnings the native press may give.*'

We are persuaded, Sir, that the part you have taken in carrying through the press the translation of the *Nil Darpan* has been in perfect accordance with your cherished convictions as to the importance of enlightening the *European* mind here on the contents of the *Vernacular Press*, and we have therefore observed with pain and sorrow the bitter personal controversy in the newspapers to which your laudable efforts in this direction have given rise.

That the *Nil Darpan is a genuine expression of Native feeling on the subject of Indigo Planting we can with confidence certify*. We are aware that there are passages in the original put into the mouths of females and others, which may grate on the ears of men of cultivated taste, but such passages only express the thoughts and ideas *current in the order of society painted in the work*. If, however, an occasional indelicacy of expression should be a reason for the suppression of a work of fiction, we fear the most ancient and best classics of our land, which are so justly valued all the world over, would remain sealed from public view; and, judged by the same standard, there are not a few of the master-pieces of European genius, both ancient and modern, which would suffer from the ordeal. We, however, apprehend that the open censure with which your effort has been visited is simply the result of an interested and factious opposition.

We have deemed it due to put you in possession of this expression of our opinion on this important question, in the belief that it may be the means of correcting the wrong impression which we have been sorry to find entertained, viz. that the community do not consider the *Nil Darpan* as an embodiment of popular feeling, and that they do not appreciate the motives which actuated you to bring its contents to the knowledge of the European public. *Nothing could be more mistaken than this*, and we do sincerely trust and hope that this letter will remove the misapprehension so much to be lamented.

We have the honour to be, Sir,
Your most obedient servants,
(Sd.) RADHAKANT, RAJA BAHADUR,
RAJA KALI KRISHNA BAHADUR,
RAJA NARENDRA KRISHNA,
BABU RAMANATH TAGORE,
And forty-three principal Natives of Calcutta.

[A subsequent address was sent in signed by 3000 natives.]

APPENDIX III

Address of the reverend J. Long to the court before sentence was passed

MY LORD,—As the result of this trial involves *consequences* extending far beyond the sphere of Calcutta, or even of India, I beg to submit, for your Lordship's consideration, the following points referring mainly to the motives which actuated me in publishing the *Nil Darpan*:—

Tried by the mode of a *criminal* prosecution, in Court, I had no opportunity to make a personal statement to the Jury. I can only state, previous to your passing sentence, what is personal to myself as to the *motives* which actuated me to publish the *Nil Darpan*, on the grounds of my being a Missionary,—an expounder of native feeling as expressed in the native press—a friend to securing peace for Europeans in the country—and a friend to the social elevation of the natives.

My Lord, it is now more than twenty years since I came to India. During that period, I have never appeared in a Court of Justice as plaintiff or defendant; my occupations have been of a very different character, and my time has been spent chiefly among natives, engaged in vernacular teaching, in the charge of a body of Native Christians, and in the promotion of Christian vernacular literature. These pursuits, along with my interest in the rural population, called my attention to the *vernacular press* of India, its uses and defects, as well as its being an exponent of the native mind and feeling. It is in connection with the latter branch of my labours, that I appear here to-day as publisher of the *Nil Darpan,* which I edited with the view of informing Europeans of influence, of its contents, as giving native popular opinion on the Indigo question. The work (the English translation I mean) was *not got up at the suggestion of natives,* or even with their knowledge, and was not circulated among them. It was commenced at the request of others. Many of the remarks of Mr. Peterson, the Council for the prosecution, are strongly in my favour, because if, as he stated, the work was so injurious in its *vernacular* dress, was I not doing a public service by making such a work known in *English*? But not in Calcutta, where it might only *lead to more bitter controversy, and where men's interests are so concerned, that all representations would have been useless, producing irritation not conviction* I circulated it chiefly among *men of influence in England* and those connected with the British legislature, which, to the oppressed of whatever colour of country, has always afforded sympathy and redress. I have aimed for the last ten years in my *leisure hours* to be an exponent of native opinion in its bearing on the spiritual, social, and intellectual welfare of natives of this land; as, for instance, when applied to on the part of the Court of Directors seven years ago, to procure for their Library, copies of all original works in Bengali or as when lately, I sent to Oxford, by request, copies of all Bengali translations from the Sanscrit; or, when I have procured for missionaries, Government, Rajas, & c., vernacular books of all kinds—I should have been a strange person, indeed, had *my* opinions harmonised with *all the chaos of opinion, in those various publications*. Why, at the request of missionaries I have procured anti-Christian works for them, as they wished to know what was written *against* Christianity.

I am charged with slandering English women in the *Nil Darpan*. Now, waiving the point that it is only planters' wives the native author refers to—I *myself believe planters' wives are as chaste as any other females of English society in India*, and it was my impression that even putting the *worst* construction on the passage the author only referred to some *exceptional* cases, not giving them as specimens of a class of females. The view however that I and others who know oriental life, have taken of this part relating to females is, that it gives the *eastern notion of the high indelicacy of any woman who exposes her face in public, or rides out in company with a gentleman*. I have heard such remarks made of my own wife; but I treated them as a specimen of village ignorance. Sir J. Shore in his 'Notes on Indian Affairs,' states instances of a similar kind, and Lieutenant Burton, who went disguised as a pilgrim to Mecca, mentions the greatest reproach the pilgrims there made against the English was, that *they shook hands with their neighbours wives!*—I regret, however, I did not append a note of explanation to this part.

I hold in my hand the first drama ever translated, and that by an illustrious Judge of this Court—Sir W. Jones, in order to give a view of Hindu society, similar service was rendered by Horace H. Wilson, by Dr. Taylor, and various other persons. I beg to say I was far from wishing to vilify planters generally, though from *sincere conviction and enquiry opposed to the Indigo system*. Thus, when summoned before the Indigo Commission, my evidence there was considered even by the planters' friends as moderate and free from invective. I was elected a member of a sub-committee of the Calcutta Missionary Conference to watch the progress of the Indigo controversy, and it was never objected then that any of my actions in connection with that Conference on this subject, were for the purpose of vilifying. I have never lived near planters, nor have I had any personal altercation with them, that would lead me to a *vindictive* course.

I ask, when hundreds, yea thousands of Bengali books were submitted by me during the last ten years to the *notice of Europeans of influence was the Nil Darpan to be the only exception?* And wherefore. *The ryot* was a *dumb animal* who did not know his rulers' language. And at the time of this *Nil Darpan* appearing matters on the indigo controversy were assuming a *threatening* aspect; so it was important that men of influence should know that the *wound was not a surface one, but required deep probing*. Could I, as a *clergyman*, have withheld a work of this sort which indicated some of the causes of the deep-seated aversion of ryots to indigo cultivation? This work, the *Nil Darpan*, was sent to me as hundreds of vernacular books have been, because it is known in many quarters that I take a deep interest in vernacular literature. Here is an illustration; these two vernacular books were sent to me a few days ago from Benares—one Robinson Crusoe in Hindi, the other a Choral Book in Urdu. Almost every week I receive new vernacular books, *and I make a point of bringing them to the notice of Europeans on various grounds*. Sir F. Haliday honoured my 'Reports on the Vernacular Press' by publishing them; so did the present Government in the case of publishing my Sketch of Vernacular Literature; so did the Vernacular Literature, Religious Tract, and Christian Tract and Book Societies shew their confidence by publishing various works of mine.

I will now state the grounds why, as a clergyman opposed to war, I published the *Nil Darpan*. My Lord, four years only have elapsed, since Calcutta was waiting in trembling anxiety for the result of the mutiny. Few could look with calmness on the future, while watch and ward were kept all night by the citizens. Many felt then, as I had long felt before, how *unsafe it was for the English to reside, in India in ignorance of and indifference to the current of native feeling*. The mutiny, in common with the Afghan war, has showed the English in India were generally unacquainted with it; so, a short time previous to the mutiny, the Sonthal war burst out unexpectedly to the public. For a long period were not thuggee and torture prevailing in India, without the English knowing any thing of them? Had I, as a missionary, previous to the mutiny, been able to submit to men of influence a native drama, which would have thrown light on the views of sepoys and native chiefs how valuable might the circulation of such a drama have proved, although it might have censured severely the treatment of natives by Europeans; the indifference of sepoy officers generally towards their men; and the policy of

Government to Native States! Such a drama might have helped to save millions of money and torrents of human blood. In Cabul, the authorities, through a false security, founded on ignorance of native opinion, entailed a loss of fifteen millions sterling on the State and the damage of England's prestige. *Had Calcutta forgotten the lessons taught by the mutiny?* I ask, was it very malicious to *reveal to the governing race the latent current of native thought* and feeling on the subject of indigo, which was convulsing the whole country, and threatening it with anarchy, incendiarism, and assassination? Would I have been justified to withhold contributing my mite at such a crisis to the great object of rousing men of influence by shewing them, from a *native* source, that the dissatisfaction was deep-seated, and that the wound must be thoroughly probed before healing measures could be efficacious?

My Lord, the *mutiny* has passed away; who knows what is in the future? And as a clergyman and a friend to the peaceable residence of my countrymen in India, I beg to state the following as a motive for my editing such works as the *Nil Darpan*. I, for years, have not been able to shut my eyes to what many able men see looming in the distance. It may be distant, or it may be near; but *Russia and Russian influence are rapidly approaching the frontiers of India*. Her influence, so manifest in Cabul twenty years ago, as shewn in a recent Parliamentary Blue Book, was beginning to be felt in India during the last mutiny. Now she goes on the principle of *divide et impera*; previous to invasion she gains over the *native population* in various countries to her side. Could I, then, as a clergyman have watched with apathy measures like those in connection with the indigo system which were *furthering this Russian policy*, and which might lead to *war and dissensions* that would retard for a long period the progress of religion, education and peaceful commerce; I now speak merely my honest convictions on this point; and I ask *if* this conviction has any foundation in reality as also *if* there be any ground for another as deeply rooted in my mind;—that *mere armies* can no more secure the English in India than they established the *Austrians in Italy* would it not be my duty as a *clergyman* to help the good cause of peace, by shewing that the great work of peace in India could be best secured by the *contentment of the native population, obtainable only by listening to their complaints as made known by the native press and by other channels.* I pass over French views in the East, but I say, *forewarned is forearmed*, and even at the expense of *wounding their feelings* in order to *secure their safety*, I wish to see the attention of my countrymen directed to this important subject.

As a *missionary*, I have a deep interest in seeing the faults of my countrymen corrected; for after a residence of twenty years in India, I must bear this testimony—that, of all the obstacles to the spread of Christianity in India, one of the greatest is the *irreligious conduct of many of my own countrymen*. Thousands of natives have said to me, 'We judge of the Christian religion by *what* your countrymen *do*, not by what they *say*; by the *life*, not by the *doctrine*.' For twenty years I have, as a missionary, been in close and confidential intercourse with natives of all classes. Often and often has my spirit been harrowed and almost crushed by a close view of the condition of the ryot, his wants and his sufferings; shut out from that ability to read, without which the pages of inspiration are locked up to him. I can see in the improvement of his *social condition a means of enabling him to enjoy the light of knowledge*. I have circulated many pamphlets in England, on 'The ryot, his teachers, and tortures,' and on the evils resulting from the ryots not having a sound vernacular education. When I have not shrunk from exposing many social evils to which the ryot is subject, I beg to ask, *could I have avoided, in my position exposing his sufferings for resisting the indigo system?*

The Chief justice here stopped Mr. Long, stating that the Court were willing to hear anything that he had to address to them in his defence. That it was not the length of the matter he was now reading, but its substance they objected to, as irrelevant. The remainder we give as from the MS. prepared by Mr. Long:—

Influential men in England have deeply sympathised with me. on these points, and have said 'You and others that expose those recesses of human suffering and degradation must let us know the results,' and I have been, my Lord, amongst those masses for years, and hope, as long as I live, and have a brain to think and a pen to write, to advocate the social elevation of the masses as incidental with the progress of mental and moral light. Should I not have been a traitor to the religion I professed, whose great Founder's motto is, 'The poor have the Gospel preached to them,' had I not availed myself of all legitimate opportunities to bring the wants and sufferings of the ryots, and the feelings and views of natives generally to the notice of men who had the power of remedying them? It may be called too *political* a course, as some now unduly restrict that term; but Christianity itself is political in the extended sense; for in the early ages it assailed the slavery of the Roman Empire; in the middle ages it afforded an asylum to the serfs against the oppressions of the feudal chiefs; at the period of the Reformation it brought freedom to the peasant's home; and in modern days it has abolished slavery in the West Indies; it has protested against American slavery, and is now throwing its mantle of protection round the aboriginal tribes throughout the world. In 1856, I delivered an address in Calcutta to the friends of Missions on 'Peasant degradation, an obstacle to Gospel propagation.' No one then objected to that address on the ground of its being political.

My Lord, I am sustained in this course by the convictions of, I trust an enlightened conscience, and confidently relying on the continued sympathy of many friends both among the European and native community, and of all in India and Great Britain, who desire to see India governed, not merely for the advantage of its *fluctuating population from Europe,* but for the benefit of, and with considerate regard for, the feelings and interests of the 180,000,000 natives over whom stretches the *Ægis* of the Queen and Parliament, I know I shall have the sympathy of good men the friends of the natives in India and in England, and of all those throughout the world who believe in the indissoluble connection of spiritual and intellectual improvement.

My Lord, a Court of *Law* has decided that the work is a libel, and it is my duty to submit to that verdict and *to act accordingly*. My conscience convicts me however, of *no moral* offence or of *any offence deserving the language used in your Lordship's charge to the Jury*. But I dread the effects of this president. This work being a libel, then the exposure of *any social evil*, of caste, of polygamy, of Kulin Brahminism, of the opium trade, and of any other evils which are supported by *the interests of classes of men*, may be treated as libels too, and thus the great work of moral, social and religious reformation may be checked.

My remarks are ended, my Lord.

Then followed Sir Mordaunt Well's sentence in the following terms:—

The sentence of the Court is, that you pay a fine of Rupees 1,000 to our Sovereign Lady, the Queen, and that you be imprisoned in the Common Jail of Calcutta for the period of one Calender month—and that you be further imprisoned till the fine is paid.

In the course of his charge to the Jury, Sir Mordaunt thought it not unbecoming to comment on the address which we have quoted to Mr. Long, though it was *in no way before the Court,* and in his judgment, he could not refrain from an *attack on the Government, without waiting to hear what it had to allege in explanation.*

The fine imposed on Mr. Long was at once paid by Baboo Kali Prosono Singh, and we hear on good authority that there were many other native gentlemen in Court anxious to be allowed to render that service to the defendant.

The next day Mr. Long was applied to, to ascertain if he would consent to a petition being sent to Government by the native community for a remission of his imprisonment, but he stated that it might embarras Government and that he should prefer the matter being allowed to rest. And this was his wisest and most proper course. It is well to suffer as a Christian; 'for this is thankworthy, if a man for conscience toward God endure grief, suffering wrongfully.' (I Peter ii. 19.)

The amount of native sympathy this case has evoked is very remarkable. Mr. Long is well known to the native community as having been engaged for many years in various works for their benefit, and as a man animated by a warm desire to ameliorate and to elevate the condition of all classes. It is most shamefully imputed to him, day after day, by one of the papers, that he is a retailer of obscene native publications, whereas the fact is, that he has for several years been an indefatigable member of the

Vernacular Literature Committee that was established by Mr. Drinkwater Bethune, Mr. John Colvin, Mr. Hodgson Pratt, Mr. Wylie, Mr. Samuels, Mr. Townsend and others, expressly to purify the native literature, and which has succeeded in putting into circulation many popular works of the soundest character. Mr. Long's 'Statement' on this subject ought to silence his assailants. It was he, as he in that statement shows, who applied for and obtained the enactment of the law for the suppression of the sale of those vile works that were once so commonly hawked about the streets of Calcutta.

But the policy of the planters and their organs in this whole controversy, seems to have been to *overwhelm every one who has resisted them with public odium,* and Mr. Long has been a chief object of their malice.

We shall not indulge in many further comments on this case, but there are some other points that call for observation.

Mr. Long himself stands before his brethren as he ever has done—a bold, faithful, benevolent man; most enthusiastically engaged in promoting the welfare of the people of Bengal, according to his own view of their wants; and most sincere, upright, and faithful in all his Missionary labours, and in his character as a Christian minister and a Christian man. There may have been points on which we have differed with him. We may have sometimes doubted the accuracy of his information and the soundness of his judgment. But there have been many other occasions on which we have admired his intuitive perception of great principles, his untiring zeal, his single-hearted labours, and his sterling worth. He is well known beyond the limits of Calcutta, and his name will be well remembered, and will be honoured ever here in this scene of his present punishment, when all the excitement of this Indigo controversy has passed away and is forgotten.

Next, as to the effect of this prosecution; the Planters think they have *gained a victory*. They may be assured that nothing has occurred, since their system was first commenced in this country, more calculated to destroy their influence. It is useless to look for a *dispassionate* judgment on the question from any in *this city*, who are involved in the vortex of discussion, or from any whose interests are at stake. But the spirit manifested in this prosecution will be noticed out of Calcutta. In other parts of India, in

Great Britain and in the United States, Mr. Long's case will be taken as involving much more than the Planters perhaps intended. It will look like an attempt to supress all freedom of discussion. It will prove the weakness of the Planters' cause. *It will arouse suspicion that Anglo-Saxon dominance, for which some so zealously contend, would mean nothing less than arbitrary government for the benefit of the European settlers, with a total indifference to the sentiments and feelings of the great native population*. Mr. Peperson in his speech for the prosecution made a most unfortunate allusion. He warned the Missionaries (whom he denounced as mischief-makers) by the example of Mr. Smith of Demerara! He instanced him as one who had *incited* the slaves to disaffection, and had *suffered the extreme penalty of the law!* He was strangely forgetful of facts! The case of Mr. Smith is a warning indeed, but it is a warning *against colonial persecution*. That man was falsely charged, and was cruelly treated by the Planters of Demerara, and he died in prison; but very soon the walls of the House of Commons resounded with the eloquence of Sir James Mackintosh and Henry Brougham, the spirit of England was aroused, and nothing probably contributed so much to the downfall of West Indian Slavery as that very case.

We await with confidence the verdict of the English people on this prosecution too. We have no fear of the result. The cultivation of Indigo, if it is to be maintained at all, must be maintained by fair prices and fair dealing. We hear as yet but little of better prices in Lower-Bengal, or of any conciliatory policy towards the ryots. The object of the Planters' Deputation to the Governor-General in February, apparently was to obtain a *Proclamation of Martial Law*.

[The Calcutta Christian Observer]

APPENDIX IV

Extract from letter addressed by H. Bell, Undersecretary to the Government of Bengal, to W.F. Fergusson, Secretary to the Landholders' and Commercial Association of British India (No. 1426 A, 3.6.1861). Delhi: Indian National Archives *Home Dept. Publ. Branch 1861 A. 28th June Nos. 103–105 & K.W. Relative to the Publication and Circulation of an English Translation of a Bengalee Play called 'Nil Durpun'.*

... The Lieutenant-Governor finds that copies of an English translation of the Publication named in your letter were sent to certain individuals.... This occurred ... not by his order. The Publication in question, however, he finds, is no libel, and does not, so far as he is aware, infringe the Law. It is an English translation of a Work of fiction; an original Bengallee Drama of a popular order, in which obviously all the dramatis personae are imaginary, and the story is manifestly a fable. It has no interest, but is an indication of strong popular feeling; as such, however, it has significance. It does not appear to the Lieutenant-Governor that even the original Bengallee Drama, judging from the translation, is likely, as the Association supposes, to be of a tendency to excite any class of persons to sedition, or breaches of the peace. In this respect, however, it is obvious that it is Vernacular Plays, not English translations, that may be dangerous, nor can the knowledge of respectable official or other European Gentlemen of the existence of such indications of popular feelings as this, be any thing but a security against actual sedition and breaches of the peace.

... It will be found that Indigo Planters, on whose behalf complaint is made, are by no means the only class, Native or European, criticised in this Bengallee Play. Faults are as unsparingly imputed to European Magistrates, Native officials, and Native Factory Amlah, as to Indigo Planters.

Nevertheless, the Lieutenant-Governor very much regrets that ... a misunderstanding has been caused ... nothing is more earnestly desired [by the Government] than the speedy introduction of a sound system for the future, such as shall carry Native feeling along with it.

Customs House Wharf.

The Charak Festival. Solvyns, colour engraving, 1808.

Thakur dalans of Nabakrishna Deb's old house.
Photograph: D.K. Lahiri Choudhury

Ardhendhushekhar Mustafi

Shivnath Shastri

Dinabandhu Mitra
Author of *Neel Darpan*, 1860

Amritalal Basu

Indigo factory, location unknown, *c.* 1900
Photographer: Bourne and Shepherd.
Courtesy: Oriental and India Office Collections, London.

Sorting of Indigo, location unknown, *c*. 1900.
Photographers: Bourne and Shepherd.
Courtesy: Oriental and India Office Collections, London.

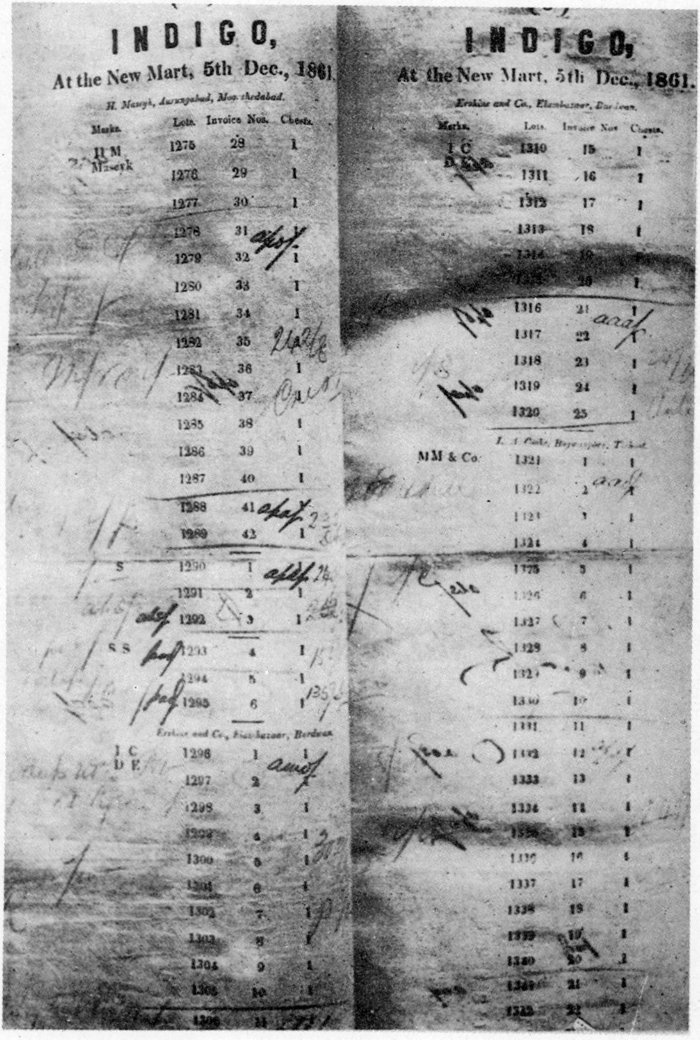

INDIGO,

At the New Mart, 5th Dec., 1861.

H. Mauyh, Aurungabad, Moorshedabad.

Marks.	Lots.	Invoice Nos.	Chests.
H M Maserk	1275	28	1
	1276	29	1
	1277	30	1
	1278	31	1
	1279	32	1
	1280	33	1
	1281	34	1
	1282	35	1
	1283	36	1
	1284	37	1
	1285	38	1
	1286	39	1
	1287	40	1
	1288	41	1
	1289	42	1
S	1290	1	1
	1291	2	1
	1292	3	1
S S	1293	4	1
	1294	5	1
	1295	6	1

Erskine and Co., Elambazar, Burdwan.

Marks.	Lots.	Invoice Nos.	Chests.
I C D E	1296	1	1
	1297	2	1
	1298	3	1
	1299	4	1
	1300	5	1
	1301	6	1
	1302	7	1
	1303	8	1
	1304	9	1
	1305	10	1
	1306	11	1

INDIGO,

At the New Mart, 5th Dec., 1861.

Erskine and Co., Elambazar, Burdwan.

Marks.	Lots.	Invoice Nos.	Chests.
I C D E	1310	15	1
	1311	16	1
	1312	17	1
	1313	18	1
	1314	19	1
	1315	20	1
	1316	21	1
	1317	22	1
	1318	23	1
	1319	24	1
	1320	25	1

[illegible]

Marks.	Lots.	Invoice Nos.	Chests.
MM & Co.	1321	1	1
	1322	2	1
	1323	3	1
	1324	4	1
	1325	5	1
	1326	6	1
	1327	7	1
	1328	8	1
	1329	9	1
	1330	10	1
	1331	11	1
	1332	12	1
	1333	13	1
	1334	14	1
	1335	15	1
	1336	16	1
	1337	17	1
	1338	18	1
	1339	19	1
	1340	20	1
	1341	21	1
	1342	22	1

'Auction Chart', Nayak, in Narendra Kumar, ed., *Calcutta 200 Years: A Tollygunj Club Perspective*, Calcutta 1981
Courtesy: Tollygunj Club Calcutta.

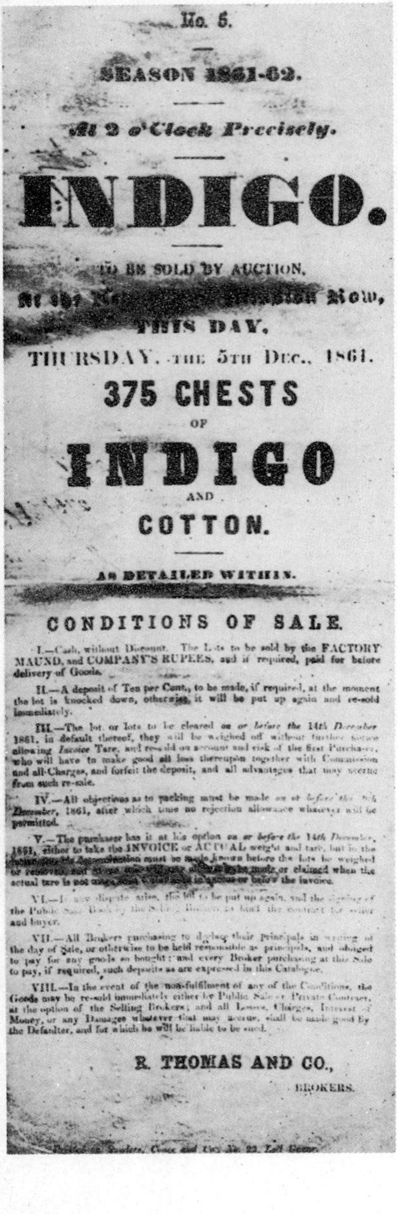

Indigo: facsimiles of R. Thomas and Company's offer, Nayak, in Narendra Kumar, ed., *Calcutta 200 Years: A Tollygunj Club Perspective*, Calcutta 1981.
Courtesy: Tollygunj Club Calcutta.

APPENDIX V

Minutes by the Honourable J.P. Grant, dated Parishnath, June 19th, 1861. Delhi: Indian National Archives *Home Dept. Publ. Branch 1861 A. 28th June Nos. 103–105 & K.W. Relative to the Publication and Circulation of an English Translation of a Bengalee Play called 'Nil Durpun'.*

I wish to place upon record, for the information of His Excellency the Governor-General in Council, the facts connected with the circulation of a translation of a Bengalee Play called Nil Darpan, with reference to a late correspondance with Mr Fergusson, the Secretary of the Landholders' and Commercial Association (lately called the Indigo Planters' Association) and to a late trial in the Supreme Court in which Mr Manuel, the Printer of that translation, was prosecuted by the Editor or Proprietor of a newspaper for libel. The words charged as libellous, I understand, are not in the Play, but a Preface by the Author which prefixed the Play.

The issue of the Pamphlet in question in any manner was not by my order or with my knowledge, and I never saw it or knew a word that was in it, until it had been circulated in the manner described by Mr Fergusson.

The issue was made through a very unfortunate error of the late Secretary. Mr Seton-Karr, several months ago, mentioned to me that he had been informed that a curious Bengalee Play had been written, the subject of which was Indigo — a genuine Native production, the translation of which might be made by a private hand, and some copies printed off at a trifling cost. I wished to see the work, partly as a curiosity, and partly because I thought it likely that it would show what the real popular feeling was on the subject better than any thing else. I thought it probable that, besides any value the Work may have as a literary curiosity, it might prove to be such that a few copies might with propriety be privately given to friends in official and private positions, with the same object as that which made me wish for an opportunity of seeing what the Work was like myself.

I have always been of opinion that, considering our state of more than semi-isolation from all classes of Native Society, Public Functionaries in India have been habitually too regardless of those depths of Native feeling, which do not show upon the surface, and too habitually careless of all those means of information which are available to us for ascertaining them. Popular songs everywhere, and, in Bengal, popular Native Plays, are amongst the most patent, and most neglected, of those means. I have always attributed our unforewarned condition when the shock of 1857 occured, to this popular defect. I did not on this occasion regard the matter as one of importance: but still the opportunity seemed a good one of knowing how Natives spoke of the Indigo question, amongst themselves, when they had no European to please or to displease by opening their minds.

Mr Seton-Karr's ideas on this point were the same as mine and I had thought that the translation and the printing of a few copies were to be a wholly private affair.

I do not believe I ever heard a word more about the matter till a copy of the printed publication was sent to me at Parishnath, without any accompanying letter....

On my first interview with the Secretary, which was before the excitement caused by the official circulation of the Pamphlet had shown itself, I found that he had been under some impression that the translation's printing was to be a Government act, paid for by Government, which impression I immediately corrected. I found also that the Pamphlet had been circulated and circulated under official frank, which was past remedy.

It is obvious, and the point is candidly admitted and deeply regretted by the late Secretary, that even upon his understanding of the affair, not a copy should have been circulated, till the Lieutenant-Governor had seen the Work and authorised its circulation. Had this error not been fallen into, unadvisedly, the Secretary's mistaken impression would have had no ill consequences.

The occurrence is extremely unfortunate and has distressed me beyond measure. It has excited irritation, when it was an object to allay irritation. But before I was made aware of it, the mischief was done.

J.P. Grant

Parishnath
19th June, 1861
(True Copy)
Ed. Whenlington,
Secy. to the Govt. of Bengal

APPENDIX VI

Letter from W.S. Seton-Karr, Legislative Member for Bengal, to E.H. Lawrence Esq., Secretary to the Government of Bengal, dated Calcutta, July 29, 1861. Delhi: Indian National Archives *Home Department Publ. Branch, 1861 A. 8th August No. 2029A, 30th July, 1861. Relative to the Publication and Circulation of an English Translation of a Bengalee Play called 'Nil Durpun'.*

As in the late trial of the Reverend Mr Long before the Supreme Court, there was mention, in the evidence, of the Bengal Office, and allusion was made to my orders as Secretary, I think it right to lay before the Lieutenant-Governor an official statement regarding my connection with the Drama called the 'Nil Darpan'.

I take this opportunity of enclosing a copy of the *Englishman Newspaper* (29th July, 1861), containing an explanation which I have thought right to make public.

The printed statement is intended to clear up charges of unfair dealing and of personal hostility to the Planters against me ...

and to explain some passages in the play itself, which I can not but think have been greatly misrepresented. What I know of the history of this Drama will be found in my printed statement, and I can only add thereto that the Reverend Long has been known to me and others as having for years devoted his attention to the Vernacular Literature of Bengal, and to the various publications, which issue from the Native Press, and circulate among the middle and lower orders to an extent of which very few Europeans are aware.

Publications on diverse social questions of interest have been by him brought to the notice of the authorities, of which a record will be found in the Bengal Office.

I believe at this point of Indian History no one will seriously dispute the paramount importance of the Government of the country being kept well informed of the state of feeling amongst Natives, as shown by their periodical and other Vernacular Literatures.

I mentioned the Work to the Lieutenant-Governor in the belief that it was my duty to bring to his notice all Native publications illustrative of popular feeling. The Lieutenant-Governor expressed a desire to see a translation of this drama, and Mr Long informed that a Native was willing to translate it — a translation was made under my sanction.

The Lieutenant-Governor approved of my noticing the work and of the act of translation and of the printing; but he never intended that so *large a number* as *500* copies should be struck off.

The copies were brought to my office, Mr Long gave me the names of several persons to whom the Work should be sent; other names were also added by me to the list ... the circulation under the official frank took place with my sanction and knowledge and without that of the Lieutenant-Governor. Copies were sent to the persons to whom copies of printed documents about the *Indigo crisis* had been sent.

I must again urge that the duty of bringing such publications as the Nil Darpan or any other such publications, to the notice of the Government is one which it behoves every officer not to neglect. ... My error lay in not examining it more carefully before issue, and in issuing it in the manner I did, without the knowledge of the Lieutenant-Governor.

I take this opportunity of placing on record the fact that, after the termination of the trial of Mr Long, I considered, looking to all that had happened, that the retention of my appointment might be a source of embarrassment to the Government.

I deemed it my duty on Thursday the 25th instant, or the day after the sentence, to place my resignation of my present appointment, as Legislative Member for Bengal, at the disposal of the Lieutenant-Governor, as well as to renounce my right to revert to my previous appointment as Secretary to Government.

I say no more on this subject, not because I do not deeply regret the general irritation which has been excited, the attacks to which the Government has been exposed and the consequences to a very earnest, zealous, and single-hearted missionary, for whom, in his misfortune, I can not but feel the deepest sympathy.

APPENDIX VII

Minute by the Honourable J.P. Grant, dated July 30, 1861, forwarding the resignation letter of W.S. Seton-Karr to the Governor-General-in-Council. Delhi: Indian National Archives *Home Department Publ. Branch, 1861 A. 8th August No. 2029A, 30th July, 1861. Relative to the Publication and Circulation of an English Translation of a Bengalee Play called 'Nil Durpun'.*

Mr. Seton-Karr's narrative correctly describes the circumstances. He candidly acknowledges the serious error committed, but I am sure that, in circulating the translation in question, he did not contemplate either giving offence to any class, or placing the Government in a false position. In putting in his resignation of his office ... he has acted the part of an honourable and high-minded man, consistently with his own established character. I do not think that the public interest would be served by acting upon the power this placed in my hands.

APPENDIX VIII

Letter from W. Grey, Esq., Secretary to the Government of India, to E.H. Lushington Esq., Secretary to the Government of Bengal, dated 8th August 1861. Delhi: Indian National Archives *Home Department Publ. Branch, 1861 A. 8th August No. 1503. Relative to the Publ. and Circulation of an English Translation of a Bengalee Play called 'Nil Durpun'.*

... The whole translation has, in the meantime, formed the subject of two trials in the Supreme Court. The printer has been fined, and the Reverend Mr Long, who superintended the translation of the Work and took part in the other proceedings, complained of by the Landholders' and Commercial Association, has been fined and imprisoned by sentence of the Supreme Court....

It appears from the published proceedings of the trial of Reverend Long, that although no copies of the translation have yet been officially furnished to the Government of India ... numerous copies had been sent, as if by the authority of the

Government of Bengal, to Members of Parliament and Public Bodies in England to some parties at a distance in the country and to the amount of no less than 20 copies to the Office of the Secretary to State in London. More than this could not have been done to give emphasis and currency to the publication of opinions endorsed by the Government of India. Nevertheless, the most important features of the apparent Sanction thus given to the circulation of the pamphlet have only become known to the Government of India through the published reports of the trial.

This course would have been a grave irregularity and in the case of a document having an important bearing on great controversial questions would not have been excusable as an inadvertence.

APPENDIX IX

Letter from: Landholders' and Commercial Association of British India

To: W. Grey, Secy. to the Government of India (no date).

Delhi: Indian National Archives *Home Department Publ. Branch, 1861 A. 15th August Nos. 63–4. Relative to the Publication and Circulation of an English Translation of a Bengalee Play called 'Nil Durpun'.*

... the Committee refrains from pressing the matter further on His Lordship's notice; although they are aware that in other cases besides that of the Nil Durpun, reports and statements have been sent home, as by Authority, and circulated to the prejudice of the Indigo Planters which was apparent from the charge made by Lord Palmerston with regard to the practice imputed to them of habitual forgery and perjury, also in the circulation of the 3rd vol. of the Indigo Selections before issue here, and in other instances.

In none of these instances were the Planters made acquainted with the charges brought against them either individually or as a body and they were thus for months deprived of any opportunity of defending themselves.

The Committee feels satisfied that after the manner in which His Lordship in Council has taken up the case, the planting interest and British Settlers will always have a fair hearing from the Government of India and their Countrymen in England....

REFERENCES

AHMAD, Qeyamuddin
1966 *Wahabi Movement*. Calcutta: FirmaK.L. Mukhopadhyay.

BADEN-POWELL, B.H.
1892 *The Land Systems of British India.* 3 Volumes. London.

BAGAL, Jogesh Chandra
1953 *Peasant Revolution in Bengal*. Calcutta: Bharati Library

BAINEBRIDGE, A.J.
1860 Letter from A.J. Bainebridge, Officiating Magistrate of Pabna, to Commissioner of Rajshahye, dated 25 August 1860. Delhi: *Indian National Archives: Home Dept. Jud. A. 79–87, 25.9.1860, Letter No. 502, 89.*

BANNERJEE, Sumanta
1989 *The Parlour and the Street. Elite and Popular Culture in 19th Century Calcutta.* Calcutta: Seagull Books.

BHADURI, Amit
1976 The Evolution of Land Relations in Eastern India under British Rule, *Indian Economic and Social History Review*, Vol. XIII(1): 45–58.

BHATTACHARYA, Neeladri
1986 Colonial State and Agrarian Society. *In* S. Bhattacharya and R. Thapar (eds.), *Situating Indian History*, pp. 106–45. Delhi: Oxford University Press.

BOSE, Kumud Behari
1903 *Indigo Planters and All About Them.* Calcutta.

BUCKLAND, C.E.
1901 Lieutenant Governor Grant (Minutes). *In. Bengal under the Lieutenant-Governors.* 2 Volumes. Calcutta: S.K. Lahiri & Co.

CANNING, Lord
1860 Correspondence of Canning to Grant. London: *India Office Library & Records: MSS Eur F 127/25, 11th Sep. 1860.*
1861 Correspondence of Canning to Wood. Delhi: *Indian National Archives: Home Dept. Judicial A. 27th February 1861, No. 45–50.*

CCO
1860 *The Calcutta Christian Observer* on the *Nil Darpan* Con-

troversy. London: *India Office Library & Records: V3210.*

CEHI
1984 *The Cambridge Economic History of India.* 2 Volumes. Delhi: Orient Longman reprint (1st edition 1982).

CHANDA, M.K.
1984 The Indigo Revolt and Christian Missionaries, *Bengal Past and Present,* Vol. CIII(I–II), No. 196–7: 46–54.

CHOUDHURY, K.N.
1965 *The English East India Company.* London: Frank Cass & Co.

CHOWDHURY, Benoy
1964 *Growth of Commercial Agriculture in Bengal (1757–1900).* Calcutta: Indian Studies Past and Present.

COOPER, Frederick and Ann L. STOLER
1989 Introduction. Tensions of Empire: Colonial Control and Visions of Rule, *American Ethnologist,* Vol. 16(4): 609–21.

DATTA, Rajat
1989 Agricultural Production, Social Participation and Domination in Late Eighteenth-Century Bengal: Towards an Alternative Explanation, *The Journal of Peasant Studies* 17(1): 68–107.

DAY, Lal Behari
1969 *Bengal Peasant Life.* Calcutta: *Eds.* Indian (repr.).

DUTT, Michael Madhusudan
1860 *Nil Darpan or The Indigo Planting Mirror.* Calcutta.

DUTT, Rajani Kant Palme
1979 *India Today.* Calcutta: Manisha Granthalaya Ltd. repr. (1st edition, London, 1940).

E.I.
1913 *Encyclopaedia of Islam: Fara'idiya.* Leiden: E.J. Brill (old edition).

E.I.C.
1818 Settlement of Europeans in India. East India Company in Parliament. London: India Office Library and Records: V3244, 1834.

FERGUSSON
1838 Letter to Fergusson (dated London, 1 September 1838). London: *India Office Library and Records: Fergusson Papers.*

FRERE, Bartle
1861 Minutes, 4 February 1861. Delhi: *Indian National Archives: Home Dept. Jud. A, 27th February 1861, No. 45–50.*

FUCHS, Stephen

1965 *Rebellious Prophets. A Study of the Messianic Movements in Indian Religions.* London: Asia Publ. House.

GB

1860 Minutes of the Government of Bengal. Delhi: *Indian National Archives: Home Dept. Judicial Branch A, 25th September 1860.*

GG

1860a Letter of the Governor-General of India to the Secretary of State for India, dated 24 December 1860. Delhi: *Indian National Archives: Home Dept. Judicial A, 27th February 1861, No. 45–50).*

GG

1860b Letter of the Governor-General of India to the Secretary of State for India, dated 29 December 1860. Delhi: *Indian National Archives: Home Dept. Judicial A, 27th February 1861, No. 45–50).*

GRANT, Colesworthy

1860 *Rural Life in Bengal, Illustrative of Anglo-Indian Suburban Life.* London: W. Thacker & Co.

GRANT, John Peter

1860 Minutes of John Peter Grant, the Lieutenant-Governor of Bengal. Delhi: *Indian National Archives: Home Dept. Judicial Branch A, 17th & 25th September 1860, No. 78–87.*

GUHA, Ranajit

1974–5 Neel Darpan: The Image of a Peasant Revolt in a Liberal Mirror, *Journal of Peasant Studies,* Vol. 2: 1–46.

1983 *Elementary Aspects of Peasant Insurgency in Colonial India.* Delhi: Oxford University Press.

HERSCHEL, William J.

1860 Statement showing the Result of the Operation of Act XI of 1860 from its Commencement upto 31st. July 1860. Delhi: *Indian National Archives: Home Dept. Judicial Part B. Proceedings, 29th September 1860, No. 69–70.*

HUNTER, W.W.

1965 *Annals of Rural Bengal.* Calcutta: Indian Studies Past and Present (repr.)

IPR

1860 *Indigo Papers by A Ryot.* London: India Office Library & Records.

KENNEDY, Paul
1987 *The Rise and Fall of the Great Powers*. New York: Random House.

KINCAID, Jamaica
1986 A Small Place, *The New Yorker*.

KLING, Blair B.
1977 *The Blue Mutiny. The Indigo Disturbances in Bengal 1859–1862*. Calcutta: Firma K.L. Mukhopadhyay.

KRISHNA, Bal
1924 *Commercial Relations between India and England 1601–1757*. London: Routledge.

KT 1267 BC
1862 Papers of the Khaspore Tahsildar. Letter from Behar, Gumastha of Khaspore Concern, to Kally Churn Gangooly, *dewan* of Mullahatty, on hearing that the Mondal was leaving the village, dated 2nd Bysack, 1267 BC. Delhi: *Indian National Archives*.

LAWRENCE, Walter
1928 *The India We Served*. London: Cassel & Co.

LUSHINGTON, George
1860 Letter from G. Lushington, Officiating Commissioner, Nadia, to the Government of Bengal, dated 15.5.1860.

MAJUMDAR, R.C.
1951–63 *The History and Culture of the Indian People. Vol. IX : British Paramountcy and Indian Renaissance*. Bombay.

MEYER
1908 *Meyers Groβes Konversations-Lexikon*. 6th Edition. Leipzig.

MISRA, B.B.
1961 *The Indian Middle Classes. Their Growth in Modern Times*. Delhi: Oxford University Press.

MISHRA, Girish
1966 Indigo Plantation and the Agrarian Relations in Champaran in the 19th Century, *Indian Economic and Social History Review*, Vol. 3(4): 332–5.

MITRA, Lalit Chandra
1903 *History of the Indigo Disturbances in Bengal with a Full Report of the Nil Darpan Case*. Calcutta.

MONTRESSOR
1862 Report to the Secretary of the Government of Bengal, dated 8 January 1862. Delhi: *Indian National Archives: Home Department Judicial A*.

MOORHOUSE, Geoffrey
1984 *Calcutta. The City Revealed.* Middlesex: Penguin (1st edition 1971).
MUKHERJEE, Nilmani
1975 *A Bengal Zamindar, Jaykrishna Mukherjee of Uttarpara and his Times 1808–1888.* Calcutta: Firma K.L. Mukhopadhyay.
MUKHERJEE, Sushil Kumar
1982 *The Story of the Calcutta Theatres, 1753–1980.* Calcutta: K.P. Bagchi & Co.
MUKHERJEE, S.N.
1970 Class, Caste and Politics in Calcutta, 1815–1838. *In* E. Leach and S.N. Mukherjee (eds), *Elites, in South Asia,* pp. 33–78. Cambridge: Cambridge University Press.
NAG CHOWDHURY-ZILLY, Aditee
1982 *The Vagrant Peasant. Agrarian Distress and Desertion in Bengal, 1770 to 1830.* Wiesbaden: F. Steiner.
NAYAK, Narendra Kumar (ed.)
1981 *Calcutta 200 Years: A Tollygunj Club Perspective.* Calcutta: Tollygunj Club Ltd.
NEALE, Walter C.
1969 Land is to Rule. *In* R.E. Frykenberg (ed.), *Land Control and Social Structure in Indian History,* pp. 3–15. Madison: University of Wisconsin Press.
PRADHAN, Sidhi and Sailesh SEN GUPTA (eds)
1958 *Nil Durpan.* Calcutta.
PSEI
1831 *Papers on Settlement of Europeans in India in 1831.* London: India Office Library and Records
RAINEY, H.L.
1873 'Jessore', *The Calcutta Review,* June 1873.
RAY, Ratnalekha
1979 *Change in Bengal Agrarian Society c. 1760–1850.* Delhi: Manohar.
1987 The Changing Fortunes of the Bengali Gentry under Colonial Rule—Pal Chaudhuris of Mahesganj, 1800–1950, *Modern Asian Studies,* Vol. 21(3): 511–19.
REID, H.M.
1860a Letter from H.M. Reid, Commissioner of Rajshahi Division, to Secretary, Government of Bengal, dated 3.7.1860. Delhi: *Indian National Archives: Home Dept. Judicial A. August 7, 1860, No. 11–16.*

1860b Letter from H.M. Reid, Commissioner of Rajshahi Division, to Secretary, Government of Bengal, dated 22 August 1860. Delhi: *Indian National Archives: Home Dept. B. No. 65.*

1860c Letter from H.M. Reid, Commissioner of Rajshahi Division, to Secretary, Government of Bengal, dated 27 August and 3 September 1860. Delhi: *Indian National Archives: Home Dept. Judicial B. Proceedings No. 87 and No. 4–5.*

RIC

1860 *Report of the Indigo Commission.* Delhi: *Indian National Archives: Home Dept. Jud. A, June 18, 1860, No. 36–46.*

RITAMBARI DEVI

1989 *Indian Mutiny: 1857 in Bihar.* Delhi: Orient Publs.

ROY, Rammohun

1831 Accounts and Papers 51. Select Committee on East India Company, 1831.

1945 English Works of Roy, Part I (*Speeches and Letters*). Edited by Kalidas Nag. Calcutta.

RYOT

1860 Selections of Papers on Indigo-Cultivation in Lower Bengal.

SAMPSON, George

1970 *The Concise Cambridge History of English Literature.* Cambridge: Cambridge University Press (3rd edition).

SEN, Sukumar

1952–56 *Bangla Sahityer Itihas.* Burdwan: Burdwan Sahitya Sabha (3rd edition).

1960 *History of Bengali Literature.* Delhi: Sahitya Academi.

SETON-KARR, W.S.

1860 Letter from W.S. Seton-Karr, Secretary to the Government of Bengal, to Mssrs. Robert Watson & Co., dated 19.11.1860. Delhi: *Indian National Archives: Home Dept. Judicial A Proc. 26–28.*

1899 *Grant of Rothiemurchus.* London.

SIM, J.D.

1860 Letter from J.D. Sim, Secretary to the Board of Revenue, to J.D. Bourdillon, Secretary to Government, Revenue Dept., Madras, dated 12 September 1860. London: *India Office Library & Records: V/26/624/2.*

SINGH, S.B.

1966 *European Agency Houses in Bengal 1783–1833* Calcutta.

SINHA, J.C.

1927 *Economic Annals of Bengal.* London: Macmillan & Co.

SINHA, Narendra Kumar
1956 *The Economic History of Bengal from Plassey to Permanent Settlement.* 2 Volumes. Calcutta: Firma K.L. Mukhopadhyay.
SPEAR, Percival
1963 *The Nabobs. A Study of the Social Life of the English in Eighteenth Century India.* Oxford: Oxford University Press.
TARA CHAND
1967 *History of the Freedom Movement in India.* 2 Volumes. Delhi: Publications Division, Government of India.
TAVERNIER, Jean-Baptiste
1977 *Travels in India.* Delhi: Oriental Books Repr. Corp. reprint.
THOMPSON, Dorothy
1989 *Queen Victoria: Gender and Power.* London: Virago.
TOZER, H.J.
1902 *British India and its Trade.* London: Harper & Bros.
VAN DER VEER, Peter
1989 *Gods on Earth. The Management of Religious Experience and Identity in a North Indian Pilgrimage Centre.* Delhi: Oxford University Press.
WATSON, Mssrs. Robert & Co.
1859 Letter from Mssrs. Robert Watson & Co., Indigo Planters and Silk Manufacturers, to the Secretary, Government of India, dated 20 October 1859. Delhi: *Indian National Archives: Home Dept. Jud. Proc Nos. 65–67.*
WATT, George
1890 *Pamphlet on Indigo.* Calcutta.
1908 *The Commercial Products of India.* London: John Murray.
WOOD, H.
1860 Letter from H. Wood, Collector of South Arcot, dated Cuddalore, 2nd August 1860, to J.D. Bourdillon, Secretary to Government, Revenue Dept., Madras. London: *India Office Library & Records: V/26/624/2.*
YOUNG, A.R.
1860 Letter from A.R. Young, Secretary to the Government of Bengal, to the Officiating Commissioner of Nadia.
ZAMAN, M.Q.
1991 Social Structure and Process in *Char* land Settlement in the Brahmaputra-Jamuna Floodplain. *Man* (N.S.), Vol. 26 (4): 673–90.

THE BLUE DEVIL

An English translation of
Neel Darpan
by
Dinabandhu Mitra

DRAMATIS PERSONAE

MEN

GOLAK CHANDRA BASU	a petty landlord of Swarpur
NABINMADHAB BASU	his elder son
BINDUMADHAB BASU	his younger son
SADHU CHARAN GHOSH	a peasant neighbour
RAY CHARAN GHOSH	Sadhu Charan's brother, also a peasant
GOPINATH DAS	the manager of the planters' estates
I.I. WOOD	a planter
P.P. ROGUE	a planter
TORAP	a Muslim peasant and a tenant of the Basu family

The planters' overseer, the magistrate, lawyers, a clerk, the court supervisor, the deputy inspector, pandits, a jail warden, a medical officer, a kabiraj, a milkman, four children, a shepherd boy, spearmen, peons, a police constable, a candidate for a job, and a native employee on the supervisory staff of the planter's factory.

WOMEN

SABITRI	wife of Golak Chandra Basu
SAIRINDHRI	wife of Nabinmadhab Basu
SARALATA	wife of Bindumadhab Basu
REBATI	wife of Sadhu Charan Ghosh
KSHETRAMANI	Rebati's daughter
ADURI	a maid in Golak Chandra Basu's home
PADI	a candy seller

PROLOGUE

I am offering *Neel Darpan* to the numerous planters of indigo, so that they may take a look at their own faces, reflected in it. May they then wipe off those dark, shameful marks of greed and selfishness, which now stain their foreheads, and adorn them instead with feelings of concern for others, pure and spotless as sandalwood paste. I shall deem it my reward if, in return for the labour I have put in, relief comes to the peasant, and the image of England and the British Government is refurbished.

Oh indigo planters! Your barbarous behaviour has tarnished the entire community of the British, which has been adorned by noble and illustrious men like Sydney, Howard, Hall and others.

Is your lust for money so overpowering that for this trifle you should, like worms eating into a lotus, damage the unsullied reputation of the British race, earned over the years?

If only you could put an end to the terrible oppression, that dreadful instrument of yours, for amassing your enormous wealth, the poor peasants could, once again, live in peace with their families.

You are indeed well aware how dreadful it is for a poor peasant when you acquire his possessions worth a hundred *rupees* and pay him a mere ten. But bent upon making money, you never like to admit this.

You say that some of you spend money on education and at times distribute medicine. Even assuming this to be true, education provided by you is more detestable than providing for shoes after slaughtering a milch cow; and your distributing medicine is no different from pouring cream into a pitcher of deadly poison.

If you call it a dispensary when you apply a few drops of turpentine to a wound caused by your *Shyamchand*, then no doubt you do have a dispensary in every house of yours.

The two editors of the daily newspapers have been filling their columns with your praises; whatever others may be feeling, you could not possibly be happy about this, since you know why they are doing this.

What an amazing power of attraction silver has! Just for earning thirty coins detestable Judas handed over noble Jesus, that great preacher of Christianity, to brutal Pilate. What then is so surprising that at the prospect of a thousand *rupees* the two editors should throw these poor, helpless peasants into your cruel jaws?

But joys and sorrows turn as a wheel does, and it seems that the sun will rise again, bringing joy to the peasants.

Realizing that it is not good for the children to be suckled by wet nurses, the great Queen Victoria, the kind mother of her subjects, has taken them into her own lap and is suckling them herself. Gentle, wise, courageous and generous, the Honourable Canning has come as our Governor-General to punish the wicked and protect the innocent. As our Lieutenant-Governor we have the very personification of Justice, the Honourable Grant, who grieves with the peasant in his sorrows and rejoices with him in his joys. Gradually truthful, experienced and impartial Government administrators, such as Eden, Herschel and some others are blossoming in the pool of the Civil Service, like so many lotuses.

Thus it is very clear that soon these great men will bring justice. Holding the *sudarshan chakra*[1] in their own hands, they will rid the peasant of the evil demon Rahu, who has seized him and causes him unbearable misery in the form of the indigo planter.

By a certain traveller

ACT I

SCENE 1

Swarpur: Golak Chandra Basu and Sadhu Charan Ghosh, sitting on the terrace of Golak Chandra Basu's granary

SADHU CHARAN. I had told you then itself, Sir, it's no longer possible to live here, but you didn't bother to listen. A poor man's advice comes true when old.

GOLAK. Yes son, I know—but is it that easy to leave one's village? This is where we've lived for seven generations. The estate those ancestors had left us was enough; we never had to look for jobs. The rice we grow lasts the year round, it feeds our guests and meets the annual Durgapuja expenses. Our mustard gives us our oil and fetches sixty or seventy *rupees* in the market. So you see, son, this Swarpur is my gold—I have nothing at all to worry about. My land gives me rice, lentils, oil, molasses; the garden gives fresh vegetables, and fish comes from our pond. Wouldn't one's heart break to leave such a happy life? And can anybody ever really go away?

SADHU CHARAN. But, Sir, that happy life is over now! Your garden is already gone, and all that property you're on the point of losing! Alas, alas, it's not even three years since *saheb* took lease of this village, and he has already ruined it. You can't bear to look at the *mondal*'s house in Dakshinpara. What it once was—and what it has come to now. Just three years ago sixty men worked for them and were fed twice a day. They had ten ploughs and at least forty or fifty bullocks. And what a courtyard—a racecourse! Summer paddy stacked around, it looked like lotuses in bloom in our Chandan lake. And what a cattle shed—like a hill! But last year they didn't have the money even to repair it and the whole thing has tumbled down on the courtyard. Mondal's two brothers—the second and the third—were very badly beaten up last year by that bastard, because they had refused to sow indigo on their rice land. There was such a lot of trouble to get them released—the ploughs, the cattle, all had to be sold. That's why those two have left the village.

GOLAK. Hadn't the eldest brother gone to fetch them?

SADHU CHARAN. He had, but they said they'd rather beg from door to door than come back here. The eldest is now left alone, he has still two ploughs left, but most of the time these are used in the indigo fields. And he also is planning to run away. Sir, you too should get over your love for this village and leave. Last year you lost your paddy, this year it may be your honour, your self-respect.

GOLAK. Honour! Not much is left of that. They've ploughed up the land all around the pond, they will now sow indigo there. That will stop our women from going to the pond. And that bastard of a *saheb* has threatened, that if I don't sow indigo on my eastern rice fields, he'll teach Nabinmadhab the lesson of his life and move him from one lock-up to another.

SADHU CHARAN. *Baro babu* has gone to the *kuthi*, hasn't he?

GOLAK. Not willingly, the bailiff took him.

SADHU CHARAN. Our *baro babu* has a lot of courage, Sir. The other day the planter was shouting at him, 'If you don't obey the overseer, if you don't sow indigo on the land his men mark out, I'll destroy your house and push the ruins into the *Betravati*. I'll make you chew paddy in the lock-up'. And my young master replied, 'First, you pay me for last year's indigo on the fifty *bighas;* till you pay me not an inch of my land will go under indigo. Who cares for the house? I'll stake my life on it.'

GOLAK. Well, what else could he have said? You see, if we could grow paddy only on those fifty *bighas*, we'd have no worries today. At least if he pays for the indigo which we grew for him, that would be something.

Enter Nabinmadhab Basu

Well son, what happened?

NABINMADHAB. Well Sir, does a cobra ever hesitate to bite a baby lying on his mother's lap, because of the grief it will cause her?

I was very nice to him, Father, but he just wouldn't budge—he stuck to his point. He would advance only fifty *rupees* and we must agree to sow indigo on sixty *bighas*! Someday, later, he will pay us for both last year's and this year's crop.

GOLAK. If all the sixty *bighas* are put under indigo, where shall we grow other crops? We shall starve to death—that's all.

NABINMADHAB. I even said, '*Saheb*, you use our men, our ploughs, our cattle to raise your indigo, you don't have to pay us for that; all we ask is that you give us our food all the year round'. And do you know, he just mocked at me and said, 'How is that possible? You don't eat rice touched by "dirty foreigners" '!

SADHU CHARAN. These days even those who work only for their food are more fortunate than us.

GOLAK. I've practically given up ploughing, even then we have to grow indigo; there's no escape. There's no arguing with the *saheb*, he will tie you up and thrash you, so there's no way out but to sow.

NABINMADHAB. Well, Father, I'll do as you wish, but I myself would like to go to court.

Enter Aduri

ADURI. *Mathakrun* asks what's going up, do you know what time it is? When will you have bath and when will you eat? That rice must have gone cold and hard!

SADHU CHARAN. [*Getting up.*] You have to do something to help me now, Sir, I've really reached the end; if I've to raise indigo on nine *bighas* with my one and a half ploughs, we better hang up our pots and pans. I shall leave now. *Pranam*, Sir; *baro babu namaskar*.

Sadhu Charan Ghosh leaves

GOLAK. It's not God's will perhaps, that we should continue to live in this house, normally, as we used to. Go son, go and take your bath.

SCENE 2

Sadhu Charan Ghosh's house: Ray Charan Ghosh enters with his plough

RAY CHARAN. [*Puts the plough down.*] That swine of an overseer—that beast—comes rushing at me like a tiger! I thought I was gone—eaten up, terrible, terrible! He wouldn't listen, he forcibly marked out the land. If indigo eats up all the five *bighas* in Sampoltala, how the hell am I going to feed my wife and kids? I'll try again, try if he doesn't give back the land. What to do—have to leave the village.

Enter Kshetramani

Has *dada* come back?

KSHETRAMANI. Father has gone to *baro babu*'s house, he should be back any minute. Aren't you going to see auntie? What are you muttering, uncle?

RAY CHARAN. Muttering, my foot! Get me some water, will you, I'm dying of thirst. I begged and begged and begged, but would that swine listen!

Enter Sadhu Charan Ghosh; exit Kshetramani

SADHU CHARAN. You're back early, Ray Charan?

RAY CHARAN. That overseer-swine has marked out our Sampoltala land *dada*, what shall we eat? How shall we survive? And what land! Like the golden *champa* flower! Sowing just one little corner, I could have paid off the moneylender. But now what shall we eat, and how shall we feed the children? Such a big family—they'll all starve and die. We need four *ser* of rice to start the day—we shall all starve and die. It's my wretched, wretched luck—what has this cursed indigo done to us?

SADHU CHARAN. Yes, we've been living on these few *bighas*. Now if they go, why stay on here? The other couple of *bighas* are saline, nothing grows there. The ploughs also will be used for their indigo, so when shall we do our own ploughing? Don't cry brother, tomorrow itself we'll sell off the ploughs and the cattle, kick the village goodbye and run away to Basanta Babu's *zamindari*.

Enter Kshetramani and Rebati with water

Here, drink some water; don't lose heart, drink. He who has given us life, will give us our food also. Now what did you tell the overseer?

RAY CHARAN. What could I say? He went on marking out the land, it was like dragging a red-hot rake over my heart. I fell at his feet, I offered him money, but he just wouldn't listen. 'Go to your *baro babu*', he said, 'Go to him, he is your father, go to him.' I've threatened to report to the magistrate. [*Seeing the overseer at a distance*]. Look, look, that swine is coming this way, and there are two bailiffs with him, they'll now drag me to the *kuthi*, I'm sure.

Enter the overseer with two bailiffs

OVERSEER. Tie him up, tie up that wretch Ray, the swine! [*The bailiffs tie up Ray Charan Ghosh.*]

REBATI. Oh God, oh God, what's this? You Sir, why are you tying him up? This is dreadful, dreadful! [*To Sadhu Charan.*] Don't just stand there and look; run to *baro babu*'s house and ask him to come.

OVERSEER. [*To Sadhu Charan.*] Where do you think you're off to? You too are coming with us—understand? Ray isn't the chap for taking advances. It means a lot more than just putting one's thumb impression—understand? You're literate, you'll have to sign your name in the register.

SADHU CHARAN. Sir, do you call this giving advance for the indigo? Isn't it ramming the advance down one's throat? But this is my fate, it follows me wherever it go. I run away hoping to avoid the blow, but there's no escape; now that blow has fallen. Before the land was leased out, this place was like *Ram Rajya*[1], and look at it today: this luck-less fellow has become a beggar in a land ravaged by famine.

OVERSEER. [*Staring at Kshetramani—aside.*] A nice dish! The very thing for the younger *saheb*. I gave him my sister and have become the head clerk. For this, one should get something, the stuff's good; well, let's see.

REBATI. Kshetra, dear, go inside.

Exit Kshetramani

OVERSEER. Let's go Sadhu, don't make trouble, come along to the *kuthi*, quietly.

REBATI. He wanted a drink of water—Sir, have you no wife and children at home? He had just put his plough down and come home, when you tied him up. My, my, he's a growing lad, by now he'd have had two meals; can he walk that long distance to *saheb*'s *kuthi* on an empty stomach? Please, Sir, in the good name of *saheb*, let him first eat something. Alas, alas—he's always so worried about the family. The poor lad is in tears, his face is so drawn! What am I to do? Why did we ever come to this accursed place—it has made us beggars! Alas, alas we're now doomed. [*She weeps.*]

OVERSEER. You old hag, don't whine. If you want to give him some water, give it, don't waste time, otherwise I'll just take him away.

Ray Charan Ghosh drinks some water, and they all leave

Scene 3

Begunberey kuthi—veranda of the big bungalow: Enter I.I. Wood and his manager Gopinath Das

GOPINATH. Sir, how've I neglected work. You yourself have seen, Sir, I leave early every morning, and am never back till late in the afternoon; and immediately after lunch, I sit down with my accounts register on advances and work till midnight—sometimes till even after one o'clock in the morning.

WOOD. You bastard, you incompetent swine! Swarpur, Shamnagar, Shantighata—three villages, no advance! Not *Shyamchand* you set right.

GOPINATH. My lord, I am your worship's slave. It is your kindness that, from a clerk, I have become a manager; you Sir, own me—it is for you to thrash me or kill me as you like; but Sir, we have a number of powerful enemies—unless they are punished, indigo has no future here.

WOOD. How I punish them, you swine, if I not know who? My money, horses, musclemen, spearmen, so much—can't punish them? Old manager always told me who my enemies. You not see, I whipped those mother-fuckers, snatched cattle, locked up the wives? Lock up wives, then bastard people taught lesson. Rascals' mischief's, we heard nothing. And you son-of-a-bitch, you wretch, tell us nothing! You asshole is very good-for-nothing. Manager job not for *kayastha*, you moron—we kick you out and give job to *keaot* man.

GOPINATH. My lord, true, your slave is a *kayastha* by caste but in his work he's a *keaot*. I ploughed down Molla's paddy seedlings and had indigo sown there; I took forcible possession of Golak Babu's freehold orchard, which had been with the family for seven generations; their other rent-free property, which they've held since the rule of the *rajas*, is also in our possession. Even a *chamar*, or a *keaot* would have refused to do what I've done—but my luck's bad, so I get no credit even for doing all this.

WOOD. That mother-fucker Nabinmadhab wants all debts paid up. We pay him not even a brash farthing; cook up his account. That sister-fucker is very court-fan; we'll see how the bastard money get from me.

GOPINATH. My lord, that man is our biggest enemy. But for him none could have proved that we burnt down Palaspur. The bastard drafted the petition himself and briefed the lawyers so trickily that the court decided against us. Because of this bastard's tricks the old manager had been jailed for two years. I did warn him—'Don't work against the *saheb*, Nabinbabu;

after all he hasn't burnt down your house', I told him. And his reply to that Sir, was, 'I have taken a vow to protect the poor tenants. If I can save one single peasant from the tyranny of the cruel planters, I shall feel I have achieved something; and for the loss of my orchard, I'll see the manager goes to jail.' It looks as if the bastard's become a *padri*—I can't make out at all what mischief he's after now.

WOOD. You afraid—didn't I tell? You big, worthless, no-good, unfit for the job.

GOPINATH. How am I afraid, Sir? Ever since I stepped into this position, I cast aside all fear, shame, embarrassment, self-respect, honour and dignity. Killing a cow, a woman, a Brahmin and burning down houses are normal things for me—like ornaments one wears daily; and as for a term in jail, Sir, it's looming over my head.

WOOD. Don't care talks, want work.

Enter Sadhu Charan Ghosh, Ray Charan Ghosh, the overseer and bailiffs saluting

Why that bastard trussed up?

GOPINATH. My lord, this Sadhu Charan is a prominent peasant, but incited by Nabin Basu he's been damaging our indigo.

SADHU CHARAN. My lord, I've never damaged any indigo, nor am I doing so now, nor have I any power to damage. Whether I like it or not I have sown indigo and am ready to sow it this time too. But in all matters there are things one can do and things one can not. Ram an eight-finger-long charge into a half-finger-long gun barrel and it'll burst. I'm a very small tenant, and have only a plough and a half and only twenty *bighas*; if out of these nine go under indigo, I naturally feel annoyed, but my annoyance can never do you any harm Sir; it can harm only me.

GOPINATH. *Saheb* is frightened in case you put him up in your *baro babu*'s lock-up.

SADHU CHARAN. Good Sir, manager, what's the point in flogging a dead horse? Can an insignificant creature like me ever dare lock up *saheb*—such a powerful, so mighty...

GOPINATH. Enough of your high-flown eloquence, Sadhu! It sounds absurd coming from a mere peasant—it hits one like a broom stick.

WOOD. The sister-fucker's become a great pundit!

OVERSEER. He explains to the peasants the points of law and the meaning of warrants—this makes trouble for us. The bastard's brother is half-dead pushing the plough all day, while insolent we are mouthing high sounding words—'mighty, powerful'...

GOPINATH. Yes, the son of a dung-selling mother has now become a *sadar-naib*. My lord, ever since the schools have been opened in the villages, the peasants have begun to give more and more trouble.

WOOD. Must write to our Association to complain to the Government—we fight till these schools closed down.

OVERSEER. The bastard threatens to go to court.

WOOD. [*To Sadhu Charan.*] You swine, you big bastard, out of your twenty *bighas* asked you sow indigo on nine; on another nine you easily sow rice again.

GOPINATH. My lord, there's so much land lying waste that why only nine, we can easily lease out another twenty *bighas*.

SADHU CHARAN. [*Aside*] Oh God! a brewer's surety is a drunkard! [*Aloud*] Sir, if the *kuthi's* labourers, ploughs and cattle could sow indigo on the nine *bighas* which have been marked out, I could then sow rice on nine more *bighas* of my land. Indigo needs four times the care paddy does, so if I have to grow indigo on the nine *bighas* marked out, all my eleven *bighas* will lie fallow. What shall I do with more land?

WOOD. Swine, big scoundrel, you take advance, I grow indigo! Son-of-a-bitch, big bastard [*kicks him*]; once tastes *Shyamchand*, he learn [*takes down the whip hanging from the wall.*]

SADHU CHARAN. Sir, why kill a fly and soil your hand; we are...

RAY CHARAN. *Dada, dada* keep mum, sign on the paper they give. Had no bath, no food whole day—am dying of hunger.

OVERSEER. You didn't go to the court, you swine? [*Twists his ears*]

RAY CHARAN. [*Breathing hard.*] Oh mother, mother I am dying!

WOOD. Bloody nigger, thrash the bastard. [*Whips him with the* Shyamchand].

Enter Nabinmadhab Basu

RAY CHARAN. *Baro babu*, I am dying, I want a little water, they're killing me Sir, they're killing me.

NABINMADHAB. Sir, these men haven't had their bath or food, their women also haven't eaten since last night. If you whip all your peasants to death, who will grow your indigo? This Sadhu Charan here took such pains last year to raise your indigo on four *bighas* of land, now if you beat him up so badly and compel him to take further advance, he might flee the village—and that'll be your loss. Please let them go now. Tomorrow morning I myself will bring them here and they'll do whatever you ask them to.

WOOD. Mind own business! You poke nose where not wanted? Sadhu Ghosh you will do what tell—it is late for my dinner.

SADHU CHARAN. Sir, what can I say? Last year you personally had marked out four *bighas* of my best land; today, the overseer marked out the rest of the good land. The plots were selected against my wishes, the indigo also will be sown like that. I promise to raise indigo for you without taking any advance.

WOOD. Oh, my advance money not money? You swine, bastard, ungrateful wretch! [*Whips him*.]

NABINMADHAB. [*Covering Sadhu Charan's back with his palm*.] Sir, you have nearly killed this poor man, what a pity! He has a large family to support. This whipping, I'm afraid, will keep him in bed for a month and that'll mean such a lot of suffering for his people. You too have a family of your own, *saheb*, how would your wife feel, if you were dragged away just as you were sitting down to your dinner?

WOOD. Shut up, you swine, sister-fucker, bastard, beef-eater. This not Amarnagar's magistrate you run to him, complain and planter's men sent to jail—this now Indrabad's Magistrate! Rascal! Before dusk you sign advance for sixty *bighas* indigo, only then you let off; you don't, I break *Shyamchand* on your head. Ten villages not take advance because you not take.

NABINMADHAB. Oh mother earth, swallow me up, let me sink down into you! Never, in my life, have I been so humiliated. [*Sighs*]

GOPINATH. Nabinbabu, don't make things worse, go home.

NABINMADHAB. Sadhu, pray to God, the only protector of the poor.

Nabinmadhab Basu leaves

WOOD. Slave of slaves, bastard! Manager, like scoundrel office, see advance big.

Exit Wood

GOPINATH. Let's go to the office, words can't soften *saheb*'s heart, Sadhu.

Your rice had been served,
But it's covered now with ashes.
Standing by you Neel,
The God of Death.
Who can save you from his clutches.

All leave

SCENE 4

Covered terrace of Golak Basu's house, Sairindhri busy twisting a ribbon for plaiting Saralata's hair

SAIRINDHRI. This is the best ribbon I've ever twisted. Our *chhoto bou* always brings good luck; whatever I do for her turns out well. Look at these twenty strands—all twenty can fit in the cup of my hand. This ribbon is just the thing for that mass of hair—and ah, such hair! Like the tresses of Goddess Shyama! And her face, lovely as a lotus, always smiling. People say sisters-in-law never get on, but I find nothing of the kind with us—I'm so happy when we're together, and she's no different from my Bipin and she too loves me as if I were her own mother.

Enter Saralata, holding a stringbag

SARALATA. *Didi*, see if the bottom of the bag's right? I hope I haven't messed it up again.

SAIRINDHRI. [*Examining it carefully.*] This time it's really nice; but—sister, sister, what've you done here? Does yellow go with red!

SARALATA. I copied your stringbag, *didi*.

SAIRINDHRI. Is the yellow there by the red?

SARALATA. No, after the red there's green, but my green's all used up, so I put yellow instead.

SAIRINDHRI. So! You couldn't wait for the market day, eh? You, sister, are always in such a hurry—isn't there a saying? —Hari is in Brindaban, but the moment I wish to see him I can't be held back.

SARALATA. Very good! Is it my fault? What can you get at that market anyway? On the last market day *thakrun* had asked *mahashaya* to get it, but he couldn't find any.

SAIRINDHRI. All right, I'll tell him,[2] next time he writes to his brother, he should ask him to send us woollen threads of five different colours.

SARALATA. *Didi*, how many days more is it for the month to end?

SAIRINDHRI. [*Smiling*] Ah, where the ache, there the finger, eh? *thakurpo* will come home, dear, as soon as the college closes—we're counting days, eh sister? The cat's out of the bag!

SARALATA. [*Blushing*] I swear, *didi*, it isn't for that, it isn't.

SAIRINDHRI. That brother-in-law of mine is one in a million—so soft spoken, such integrity. When his brother reads out his letters, it's as if honey drips from them; I've never seen a young man so respectful to his elder brother, and the brother also has so much love for him; the very name of Bindumadhab makes his heart swell with pride (strokes Saralata's cheek fondly). My *thakurpo* and our *chhoto bou*—it isn't easy to find the likes of them. *Saralata* is just the right name for her—the very image of innocence. Now where did I leave my tobacco-pouch? I just can't do without my tobacco, but somehow I always keep misplacing that pouch!

Enter Aduri

O, Adar, could you, O *didi*, please get me my little tobacco-pouch?

ADURI. Now where am I going to look for it, tell me!

SAIRINDHRI. Listen dear, you'll find it stuck in the right side thatch of the kitchen veranda.

ADURI. Oh dear! I'll have to fetch the ladder then, how else can I reach the thatch, tell me?

SARALATA. She's understood every bit, hasn't she?

SAIRINDHRI. Funny, she understands *thakrun*'s every word. Listen Aduri, you don't know what a veranda is, do you? And of course, you don't know what a *dan* [right][3] means, eh?

ADURI. Why should I be called a *dan* [witch]? It's my bad luck! If a woman's poor and old, loses her teeth, at once she becomes a witch! I'm going to tell *mathakrun*—huh—am I that old that I can be called a witch?

SAIRINDHRI. Really impossible! [*getting up*] Sister you be here, I'll be back, I'd like you to read to me Vidyasagar's *Betal*[4].

Exit Sairindhri

ADURI. Oh, that Sagar! That man who marries off widows? Shame, shame! They say there are two groups now, I'm with the King.

SARALATA. Tell me Aduri, did your man love you?

ADURI. Young mistress, don't open up that wound. Every time I remember his face, I want to cry. He loved me so much. He wanted to buy me bangles... [*hums softly to herself*]

Are bracelets that heavy my love,
Are bracelets that heavy?
If there is someone I really love,
On her arms shall be bracelets, put by me.

'See if they fit you', he'd ask. He wouldn't let me sleep; if I dozed off, he'd say, 'My life, are you asleep?'

SARALATA. Did you call your man by his name?

ADURI. Never, never, shame, shame! One's husband is one's lord, can one ever call him by his name?

SARALATA. Then how did you call him?

ADURI. I'd say, 'Oh you', 'look here', 'listen'—

Sairindhri re-enters

SAIRINDHRI. Now, who's been teasing this crazy girl again?

ADURI. She was asking me about my man, so I was telling her.

SAIRINDHRI. [*Laughs*] Find another madcap like our *chhoto bou*! Of all things, she has to listen to the details of the story of Aduri's man!

Enter Rebati and Kshetramani

Come in, Ghosh *didi*, come in. So many times I sent word to you to come, but you couldn't make it. *Chhoto bou,* look, your Kshetramani also has come; for the last few days, she's been eating my life out—plaguing me, '*Didi*, now that Kshetramani is back from her father-in-law's house, why hasn't she come to see us?'

REBATI. How very kind you are! Kshetra, bow to your aunties. [*Kshetramani touches their feet.*]

SAIRINDHRI. [*Blessing her*] May you live long with your husband, may you always wear vermilion in your hair[5], even when it grows grey; may you have a bonny boy when you return to your father-in-law's house.

ADURI. When your mistress is with me, she goes on like a chatterbox, and now, look, not a word—afterall the girl touched your feet—say something—'live', 'die'—something!

SAIRINDHRI. God forbid, precious gift of Mother Shashti[6]—God forbid! Aduri, will you please go and ask *mathakrun* to come?

Aduri leaves

A real scatterbrain—doesn't realize what she's saying. How many months gone is Kshetra, dear?

REBATI. So far, *didi*, I haven't said a word about it to any one; and my luck is so bad, even I myself don't know for sure. But you are our own people, I'll tell you—she'll be in her fourth month when the few days of this month are over.

SARALATA. But she doesn't show it.

SAIRINDHRI. Now this is another scatterbrain! The girl isn't three months gone yet, and she's asking why her belly's not huge!

SARALATA. Kshetra, why've you taken off your tiara-ornament?

KSHETRA. Seeing me wear that, my husband's elder brother got furious, and told my mother-in-law that tiaras are only for prostitutes and rich men's daughters. Hearing that I nearly died of shame, and I took it off that very day.

SAIRINDHRI. *Chhoto bou*, it's getting dark. Get the clothes down from the terrace, dear.

Re-enter Aduri

SARALATA. Come Aduri, let's get the clothes down from the roof.

ADURI. Let your master come home first, ha, ha, ha,

Saralata sticks out her tongue, and leaves

SAIRINDHRI. [*In mock anger, laughing.*] You wretch, pulling her leg all the time! Where's *mathakrun*? Oh here she's.

Enter Sabitri

SABITRI. Ghosh *bou* you've come, I am so pleased, oh, you've brought your daughter also, that's nice; Bipin wanted to be petted, I could come only after pacifying him and sending him out.

REBATI. *Mathakrun*, let me touch your feet, Kshetra, bow to Grand-ma. [*Kshetramani touches Sabitri's feet.*]

SABITRI. May you be happy, may you be the mother of seven

sons—[*sound of coughing from inside*] *baro bou ma*[7], go to your room; I think Nabin is awake. My poor son—he has no time for bath or food—pressed down by so many worries my Nabin has grown so thin. [*Voice within 'Aduri'.*] *Ma*, go, may be he wants some water.

SAIRINDHRI. [*Aside to Aduri.*] Aduri, he's calling you.

ADURI. Yes, he's calling me, but he wants you.[8]

SAIRINDHRI. Wretch! Ghosh *didi*, do come again. [*She leaves.*]

REBATI. *Ma*, we're by ourselves now, I'm in great trouble. That sweetmeat maker Padi, turned up at my place yesterday.

SABITRI. Ram! Ram! Ram![9] Does any one ever permit such a disreputable character to enter one's house! An evil woman, she's finished, all that remains to be done is to declare her a harlot.

REBATI. Well Mother, what could I've done? There's no wall round our house. Once the men go out to the fields, it's more like a market place than a house. You know what that strumpet said?—Mother, my hair's standing on end—that harlot tells me, that the junior planter saw Kshetra as he rode by and is now mad for her, and she has to go with her to *saheb*'s room in the bungalow.

ADURI. [*Spits*] Such stench—horrible onion stench! Can we ever go near a *saheb*? Such bad smell[10] [*spits again*] stench of onions! I must never go out again alone, I can stand every thing but not the stench of onions—[*spits*] dreadful smell.

REBATI. Mother, don't poor people have their religion! Or their religion has no value, because they're poor! That harlot said, he'll give us money, release our rice land, and give a good job to my son-in-law. Damn money—can one sell one's religion? Can it ever be priced? You know Mother, if she hadn't been *saheb*'s woman, I'd have kicked her in the face and smashed it. My Kshetra is too stunned to speak, since yesterday she's been having trembling fits.

ADURI. My, my he has such a beard[11]! And when he talks, it's as if a billy goat is bleating; till he shaves off that beard and gives up eating onion, I won't go anywhere near him, [*Spits*] Ah! that horrible onion smell.

REBATI. And Mother, that terrible woman tells me, 'If you don't send her with me, she'll be kidnapped by his musclemen.'

SABITRI. This isn't a lawless land where Mags[12] rule! Under English rule no one can break into a house and carry off a woman.

REBATI. Mother, they can do anything to a peasant. If they can carry off wives to pressurize their men, to take advance for sowing indigo—can't they do that if they simply fancy a girl? You haven't heard Mother, because Nayda wouldn't agree to settle the case out of court they broke into his house and kidnapped the wife of his second son.

SABITRI. What lawlessness! How terrible! Have you told Sadhu about this?

REBATI. No Mother, this indigo business has hit him so hard, he's already losing his mind. If he hears of this, there's no knowing, he might in anger, bash his head with his axe.

SABITRI. All right, I'll ask *karta*[13] to talk to him; you don't need to tell him. How awful! The planters can do anything! Why should they say then that *sahebs* are very just? My Bindu also speaks so well of them. Aren't these planters also *sahebs*? Or are they the *chandals*[14] among them?

REBATI. That harlot sweetmeat-seller, said something else, too; may be *baro babu* hasn't heard of it yet—some new order has been passed, the *sahebs of the kuthi* and the *machertok*[15] *saheb* can join hands and send to jail anyone they want for six months. They're now planning to lay a trap to catch our karta *mahashaya*, she said.

SABITRI. [*Sighing*] If that be the will of Ma Bhagabati[16], so be it.

REBATI. She said so many things, Mother, I couldn't even understand, there's no *pill*[17] against the sentence, she said.

ADURI. May be they've given it medicine to stop pregnancy.

SABITRI. Aduri, dear, will you keep quiet?

REBATI. The *kuthi's* mistress has already written to the *machertok saheb* to start the case, and *hakim* always listens to every thing she says.

ADURI. I've seen that mistress, she has no modesty, no shame. After all he's the *machertok* of our district! So many redturbaned men, swords in hand, are around him—we tremble when we hear his name—and this lady goes riding with him on her horse! A married woman riding a horse! Keso's aunt had only smiled a little, while talking to her husband's own elder brother, and people put her to such shame—and he's afterall the *machertok* of our district!

SABITRI. You fool—you'll get us in trouble one of these days, I can see. It's getting dark now, Ghosh *bou* you better go home, Mother Durga is with us.

REBATI. I'll take leave of you then, Mother. I'll have to stop by the oil press, I'm out of oil; only then can the lamps be lit.

Exit Rebati and Kshetramani

SABITRI. [*To Aduri.*] Do you always have to put in your comments in everything?

Enter Saralata, a bundle of clothes on her head

ADURI. Here comes the washerman's wife!

SABITRI. Why should she be a washerman's wife? She's my girl of gold, my Rajalakshmi[18], looking after our welfare [*pats her back*]. Well child, was there nobody else, that you had to bring these clothes in? Can't you sit still for five minutes? Only a mad mother could've given birth to such a madcap—how did you manage to have such a big tear on your *sari*—eh? I'm sure you've bruised yourself, my poor, poor darling. My child has the bloom of a red lotus—just a bruise looks like she's bleeding. Now my little *Ma,*[19] don't rush up and down the stairs in the dark.

Enter Sairindhri

SAIRINDHRI. Let's go to the pond *chhoto bou.*

SABITRI. Yes *ma,* you two sisters better go for a dip before it gets too dark.

All leave

ACT II

SCENE 1

Begunberey bungalow's warehouse: Torap and four other peasants sitting

TORAP. Why don't they kill me? I can never be ungrateful! It's *baro babu* who has saved my caste—but for his kindness I wouldn't be on my legs today—the man protecting my cattle and ploughs. And I should go and give false evidence against such a man's father, to send him to jail? Never! I'll never do that—I'd die first.

FIRST PEASANT. Remember that rod? It doesn't bend. The blow of that *Shyamchand* is a terrible one. It isn't that we're all brazen brutes, and haven't eaten his salt—but what to do! They won't keep us in one piece if we don't give evidence; Wood *saheb* stamped on my chest—see the blood's still streaming down. The swine's feet are like the hooves of a plough-ox.

SECOND PEASANT. Those were nails, sharp nails—didn't you know *sahebs* wear nailed boots?

TORAP. [*Grinding his teeth.*] To hell with nails! That blood makes me feel sick. Ah—what shall I say—if I get hold of that swine once in the field of Bhatarmari, I shall give him such a punch, that his jaw will go flying to the skies! That'll put an end to all his 'god-dams' and the dirty words he mouths.

THIRD PEASANT. I'm a day-labourer, and keep going on the little I earn; they simply can't say that I obeyed *karta mahashaya* and refused to sow indigo—then why have they locked me up in the warehouse? She's pregnant, and the time is coming for the ceremony; I thought, well, I'd better work hard and save a little to invite a few friends and relatives—and now I'm rotting inside this lock-up for the last five days! They're sure to send us on to Andarbad.

SECOND PEASANT. I'd gone to Andarbad once, to that Bhabnapuri bungalow. You know that *saheb* of Bhabnapuri, whom everybody praises? Well, once that swine had brought a criminal case against me. I saw a lot of funny things that time, inside the court room. Bah—as soon as the *machertok* said 'Hallo', the two bastards of *mukhtars* ran to him bellowing, as if sitting near their tails he had hit them with the plough. And then they went hammer and tongs at each other. I thought the white bullock of Sad Khan and that stupid bull-calf of Jamatdar were butting each other in the field of Mainamath.

TORAP. Were you found guilty? The *saheb* of Bhabnapur wouldn't cause trouble, unless there was good reason for it—'tell the truth and you ride a horse'. Had these swine all been like that rascal, none would have spoken ill of those bastards.

SECOND PEASANT. Indeed! What nonsense! Isn't there a saying—I sought out Kelo's mother thinking she was good, but she tells me that she fucked her son-in-law. All that talk that the swine doesn't lock up people illegally is rubbish. Seven peasants were found inside his lock-up and one of them was just a kid; that dog had filled his lock-up with cows as well as calves—he was playing havoc with our lives—the swine!

TORAP. As soon as a good chap comes, they swallow him up; they're forming a committee to get rid of this *machertok*.

SECOND PEASANT. I can't understand if they found fault with the *machertok* of this district or of the other district.

TORAP. Because he didn't go to their *kuthi* for dinner. To hook him they had cooked a big feast, but he hid himself like a stolen cow, and didn't eat with them. He comes from a good family, why should he go to the house of an indigo devil? I think I know the reason—these scoundrels must be the low-caste people of England.

FIRST PEASANT. Then how come the last Governor went about feasting with the planters, like a bridegroom before his wedding day? Didn't you see how the bastards took him to the *kuthi*, in a procession, all dressed up like a bridegroom?

SECOND PEASANT. Maybe he had a share in their business.

TORAP. Absurd—how can a Governor have a share in indigo business? He wanted to be popular; if God blesses our present Governor with long life, we shall have food to eat and the indigo devil of the bastards won't be sitting on our backs.

THIRD PEASANT. [*Frightened*] I'm then finished! My *bou* says, once foreign devils catch you, they don't leave you easily.

TORAP. Ah, why did they have to put Manni's brother here? The fool can't even understand what's real and what's made up. When out of sheer terror of these *sahebs* people began to leave the village, Grandpa Basaruddin composed a verse

The cat-eyed, foolish, fatty
is the indigo-devil of the indigo-kuthi.

Grandpa Basaruddin is a poet and has written many songs.

SECOND PEASANT. Haven't you heard the one by our Netai?

Your caste is robbed by the padri,
Your rice is robbed by the indigo-monkey.

TORAP. This is a new one, 'caste robbed'? Quite true, 'caste robbed by the *padri* and rice robbed by the indigo-monkey'!

FOURTH PEASANT. Ha! Heaven knows what's happening at home; I'm a tenant in another village—when did I come to Swarpur to be told by Basu *mahashaya* not to take advance? My youngest son had jaundice, so I came to Swarpur to get some sugar candy from Basu *mahashaya*—such a kind man, and so handsome—as if Gajendragamini[1] had come down to earth and were sitting here.

TORAP. How many *bighas* have been forced on you this time?

FOURTH PEASANT. Last time it was ten *bighas*, and the trouble they gave about the payment! This time they've forced advance for fifteen *bighas* on me. I'm doing whatever they want me to, but even then they don't stop harassing me.

FIRST PEASANT. For the last two years I've been trying to bring a small piece of land under the plough. At last it was ready and I kept it for sowing sesame—but the other day the junior planter came riding, stood there and had that land marked for indigo; how can a peasant survive this way!

TORAP. All this is the mischief of that swine of a manager; the *saheb* can't keep track of all the land we have. It's that bastard—goes round and round like a mad dog, and the moment he finds a good piece of land, he stakes it out at once for the *saheb*. *Sahebs* aren't short of money, they don't have to go to money-lenders, then why should these rascals go about like this? Grow indigo if you want to, buy bulls, cows, order ploughs. If you can't plough yourself employ labourers, you've enough land; plough up the whole village if that's what you want, we don't mind working together. In a couple of years then, there'll be such a surplus of indigo. But these bastards won't do that—they love the flesh of the dead and find the blood of the peasant sweet, so they go on and on sucking. [*Within: Growns—Oh! Oh! Oh! ma! ma! ma!*] Ghazisaheb, Ghazisaheb, Shrine! O Holy Shrine—you all chant God's name—inside there's a ghost—sh! sh! Don't say a word.

Voice within. Oh indigo, indigo, did you come to this land to destroy us! Ah, ah I can't stand this pain any longer! I don't know how many more *kuthis* this concern has. In six weeks I've been inside fourteen lock-ups, I don't know where I'm now and how can I know—they've been moving me around from place to place, always at night, always blindfolded—oh Mother, Mother where are you?

THIRD PEASANT. Ram! Ram! Ram! Kali! Kali! Durga! Ganesh! O demons!

TORAP. Sh!—keep quiet.

Voice within. Ah, ah, I can get out of this hell only if I take advance at the rate of five *bighas*; uncle! uncle! we must take the advance, but how to send word—I'm near death, I haven't the strength of talk; oh mother, it's now six weeks since I last saw you.

THIRD PEASANT. You heard, even after death, even as a ghost, one can't get out of the clutches of advance—I must tell this to my wife.

FIRST PEASANT. You really are such an ass!

TORAP. From his talk, he seems to be from a respectable family. Uncle Paran, could you lift me up on to your shoulders? Then I could quickly ask him through the chink where he comes from.

FIRST PEASANT. But Torap, you're a Muslim....[2]

TORAP. Then you stand on my shoulders and take a look. [*He bends down.*] Get up. [*First peasant gets up.*] Hold on to the wall, take your face close to the chink. [*Suddenly sees Gopinath at a distance, coming up.*] Get off uncle, be quick, get off, that Gupe swine's coming this way. [*First peasant jumps down.*]

Enter Gopinath Das and junior planter Rob Rogue, the whip Ramkanth *in his hand*

THIRD PEASANT. Manager, Sir, there's a ghost inside that room, it has been crying all the time.

GOPINATH. If you don't say what I ask you to say, you also will become a ghost like that. [*Aside to Rogue.*] They've sensed something about Majumdar; he must be shifted, it was a mistake to have kept him here.

ROGUE. All that later—which scoundrel's unwilling? Where's that wicked bastard? [*He stamps on the ground.*]

GOPINATH. All these have come round—only this swine of a Muslim, this bastard says he can't be ungrateful.

TORAP. [*Aside, to himself.*] My God! What a whip! Now I'll say 'yes' to every thing. I know what I'll do then. [*Aloud*] Good Sir, I'm willing too.

ROGUE. Shut up, you son of a pig, you'll find *Ramkanth* sweet [*He whips and kicks him.*]

TORAP. Allah! Allah! Oh mother, I'm dying. Uncle Paran, give me a little water, I'm dying of thirst—oh father, father.

ROGUE. I'll piss in to your mouth! [*Kicks him.*]

TORAP. Good *saheb*, good *saheb*, I'll do whatever you ask me to, I vow by God.

ROGUE. Sister-fucker's coming round. Tonight send them all for trial; write to lawyer. Their evidence will be recorded first, then only they go out. Court clerk should go with them. [*To third peasant.*] Why are you crying, you fool? [*Kicks him.*]

THIRD PEASANT. Oh wife, where are you? They're killing me—oh mother, oh wife, oh mother, they've finished me, I'm dying [*He drops down.*]

ROGUE. Bastard! Is mad!

Rob Rogue leaves

GOPINATH. So, Torap, you had to taste both—the whip and the boots!

TORAP. Mr manager, Sir, give me a little water to drink, I'm near death.

GOPINATH. Yes, this is an indigo lock-up—a potsherd hut—you go on sweating and go on drinking; come along, all of you, I'd better give you some water to drink.

All leave

SCENE 2

Bindumadhab's bedroom: Saralata reading a letter

SARALATA. Her beloved hasn't come, the young woman's like a lotus crushed by an elephant. I feel so terribly disappointed; no skylark waits for fresh rain with as much longing as I did for the arrival of my love. What *didi* said was true, I was counting days. One day was like one year. [*Sighing*] now there's no hope at all of his coming. But he must succeed in the noble task he has taken upon himself; only then will he be happy. Dearest, we are born women; even if we are five girls together, we can't go on our own to the park, we can't go for a walk in the town, it isn't possible for us to start any welfare society; for us there's no college, no court, no Brahmosamaj[3] to go to. When a woman is sad, there's nothing to take her mind off; how can I blame my heart if it's disconsolate. Our only shelter is our husband; he's the centre of our thoughts, our consciousness, our perception; he's our wealth, our club, our society. A husband is the only treasure of every chaste woman. Oh letter! you've come from my dearest love; let me kiss you. [*She kisses the letter.*]. You carry the name of my beloved, let me press you to my burning heart. [*She presses it to her bosom.*] Ah! —how sweet are his words; the more I read it, the more elated I feel, let me read it once again. [*She reads the letter.*]

Dearest Sarala,
No letter can tell how much I have been longing to see that sweet face of yours. My joy knows no bounds when I press your lovely face to my heart. I was hoping that the days of my happiness were drawing near, but my joy has turned to sadness. The college is closed, but I am in great trouble; if it is God's will that I should not succeed, I shall not be able to show my face. The indigo planters have brought a false case against Father. On the sly, they are trying their best to get him imprisoned. I have already written to brother about the matter. I am staying on here to look after things. Please don't worry, by the grace of God, we shall overcome all this.

Dearest, I have not forgotten that you wanted a Bengali translation of Shakespear. Just now it is not available, but my friend Bankim has lent me his copy, and I shall bring it with me when I come home. Darling, it is such happiness to be able to read and write: though I am so far from you, I can talk to you, and had Mother not objected to your writing, I could—like the *chakor* bird—have drunk in deep the nectar of your writing and attained fulfillment.

Yours ever,
Bindumadhab

Yes, you're mine, I know that for certain. Dearest, if ever you are blemished, where then shall we look for an ideal character? By nature I'm restless, I can't sit still in one place for five minutes—*thakrun* teases me and says that I'm a mad woman's daughter; but where's all that resilience gone? I've been sitting here for hours in the same place, holding this letter from my darling. All that restlessness has now entered into my heart—when rice boils over, the top settles down, but inside, the rice goes on boiling; it's the same with me. I don't even feel like laughing—laughter is wife of happiness—when happiness dies, laughter dies with it. Dearest, if you succeed, every thing will be all right; the thought of your sad face plunges me into darkness. Oh foolish heart, won't you be quiet? Your restlessness you can hide, none can see you cry, none can hear you cry. But you, my eyes! You will shame me. [*Wipes her eyes.*] I can't leave the room till you're quiet.

Enter Aduri

ADURI. What are you doing? Your sister-in-law's waiting for you to go to the pond—what has happened? All of you're looking so grim!

SARALATA. [*Sighing*] Come, let's go.

ADURI. You haven't even oiled your hair! The tips are getting so brown! Even the letter you haven't put aside! Our younger master asks about me in all his letters!

SARALATA. Has *baro thakur* finished his bath?

ADURI. He has left for town; there's a case on in the court—doesn't your letter say that—*karta mahashaya*'s in tears.

SARALATA. [*Aside*] Yes my dearest, unless you succeed, you won't be able to show your face any more. [*Aloud*] Come, let's go to the kitchen and oil my hair.

Both leave

SCENE 3

Swarpur, a road junction: enter the sweetmeat-seller Padi

PADI. That son of a bitch, that *amin*'s ruining this village. Do I like it? Getting young girls for *saheb* and digging my own grave? Ray would have showered such a feast of blows on me the other day that, but for Sadhu *dada*'s holding him back, all would have been over with me. But really, I feel so utterly miserable when I see Kshetramani's face; haven't I got a heart, simply because I've a lover? Whenever she sees me, she comes and sits by me, calls me 'Auntie'—a golden fawn that—can a mother ever throw her to a tiger? That junior planter's lust has no limit. I'm there, that tribal woman Kali's there, but, ah—such an awful shame! For money I've sold away my caste, my family. I had to touch the bed of that tribal woman! The senior planter has threatened to thrash me if he ever finds me; he says he'll cut off my nose and ears. That knave's senile, that rascal, that harlot's bed-fellow, he can lock up women, can kick them in their buttocks, but the swine's not like that other fellow. Let me see, I'd better tell that shameless *amin* that I can't do this. I can't even go out these days to the village. Those village urchins, those louts, sons of whores come after me, like crows after a sparrow.... [*Within. a song*]

When sitting in the field, the corn I reap,
Those eyes of love, at me they peep.

Enter a cowherd

COWHERD. *Saheb*, is it true that your indigo seedlings have been eaten up by pests?

PADI. May pests eat up your mother and sisters, may Yama[4] sieze you and drag you to Kalmighata, may your mother be childless!

COWHERD. I've ordered the blacksmith to make a pair of rakes....

Enter a muscleman of the planter

Oh God! It's the *kuthi's* muscleman! [*He runs away as fast as he can.*]

MUSCLEMAN. O Lotus-faced! By your grace, the price of tooth powder is soaring high.

PADI. [*Looking at the muscleman's belt.*] Your belt looks like a necklace.

MUSCLEMAN. You know sweetheart, between a bailiff's uniform and a dancer's costume—well, there's no difference.

PADI. I'd asked you for a black calf, you never gave it, I'll never again ask you for anything.

MUSCLEMAN. Lotus-faced, don't be so peppery. Tomorrow we're going to raid Shyamnagar. If I find a black calf there, you'll find it tethered inside your cowshed. I'll see you in your shop on my way back from the fish market.

Muscleman leaves

PADI. Raid, raid—these *sahebs* have nothing else to do! Only if they were a little less brutal—peasants could breathe again and you too would get your indigo. The poor clerk of Shyamnagar begged and begged to get his ten fields back, but does a chief read scriptures! The senior *saheb*—that rascal—sat tight, as if his tongue were singed.

Enter four school children

CHILDREN. [*They put away their bundles of palm leaves, clap and sing.*]

Oh dear candy seller, candy seller have you
steeped the indigo, the indigo, - - -
oh dear candy seller, my candy seller - - -

PADI. Keshab, my good son, shame, shame, should you say such things to auntie?

CHILDREN. [*Dancing*]

Candy seller sweet—it's late
Haven't you steeped the indigo yet?

PADI. Sh, sh—Ambike, my little *dada*, you mustn't say these things to *didi*!

CHILDREN. [*Dancing round her.*]

Oh candy seller friend, have you
steeped the indigo?
Oh candy seller, my friend, have you
steeped the indigo?
Oh candy seller, my friend - - -

Enter Nabinmadhab Basu

PADI. Oh God—what a shame—what a shame—I've shown my face to *baro babu*! [*She rushes out, pulling her sari over her head.*]

NABINMADHAB. Wicked creature, evil thing. [*To children.*] You're again playing in the street? It's very late, go home now. [*The children leave.*] If only this indigo torture would be over! In five days, I could set up a school for these kids, the Inspector of Schools is a very nice man, education makes all the difference; though young, he's very wise, and is keen to have a school here, and I don't mind at all spending on such welfare work. Actually our big thatched hall would make a lovely school; can there be greater pleasure than to see children coming to my own house to study—that is the reward for which one earns and works hard. Bindumadhab had brought this gentleman along with him, and he was keen that the village people should co-operate in the founding of the school. But after seeing the terrible condition here, he kept quiet. Our Bindu has a lot of understanding, and by nature he's so gentle and courteous. For someone so youthful his wisdom is amazing—it's like a fruit borne by some young plant. His letter's so sad, that it would break the heart of a stone—may be the planter also will relent. How can I now face home? I don't see any way out. Of those five not one could be traced. Heaven knows where they've been whisked away to—none could tell. Torap, I think will never tell a lie, but if all the other four stand as witness that'll be a disaster, and I haven't managed to get a single man on our side. On top of this the magistrate is a great friend of Wood saheb.

Enter a peasant, two peons of the court, and some servants of the planter

PEASANT. *Baro babu*, please look after my two kids; there's none to feed them. Last year I gave them eight cart loads of indigo, not a pice have I been paid for that. Now they charge me with arrears, and I'm in handcuffs and will be taken to Andarbad.

SERVANT. Advance for indigo is like the washerman's mark; once it's there it doesn't come off. Come along, rascal, we'll have to go *via* the manager. Your *baro babu*'s going to meet the same fate soon.

PEASANT. Yes, yes, I'm coming, I'm not afraid, I'd rather rot in jail than sow indigo for that bastard. Oh God, Oh God, there's none to help the poor. [*He weeps.*] *Baro babu*, give my two children something to eat, they took me straight from the field, I couldn't even see them once.

All leave, except Nabinmadhab Basu

NABINMADHAB. What dreadful injustice! These two kids will starve to death, like the younglings of a rabbit caught in a snare.

Enter Ray Charan Ghosh

RAY CHARAN. If *dada* hadn't held me back, I'd have cornered that whore and killed her—that harlot. Then, or after six months they'd have hanged me—matters little.

NABINMADHAB. Oh Ray Charan, where are you off to?

RAY CHARAN. Mistress has asked me to call the priest, that bitch Padi says tomorrow peons'll come with court summons.

Ray Charan Ghosh leaves

NABINMADHAB. Oh God! That such things could ever happen to us! Father's such a simple man, so straight, so generous—he never had anything to do with squabbles or confrontations, he never likes to step out of the village. The very mention of criminal courts frightens him. He wept after reading the letter. He'll lose his mind if he has to go to Indrabad, and if he's sentenced he'll drown himself. God, God, I'm still alive, yet Father has to suffer so much; Mother hasn't lost hope. She's brave, not nervous like Father. She's been praying to Goddess Bhagabati[5] from the bottom of her heart. And in this forest fire my lovely doe-eyed wife is my companion. She's deeply upset that I might die in the same manner as her father had died, in the planter's lock-up. How many can I comfort! Shall we all flee the village, or should we stay on and try to help others as a part of one's duty? I haven't been of any use to the Shyamnagar people; still nothing's impossible, if one really tries. Let me see what I can do.

Enter two teachers

FIRST TEACHER. [*To Nabin*] Sir, we have heard that Golak Chandra Basu lives in this village. My uncle told me that he is a very good man and is highly respected among the *kayasthas*.

NABINMADHAB. [*Bowing*] Respected sir, I am his eldest son.

FIRST TEACHER. Really, really, how nice—how very nice—such a son is a reward not simply for ordinary, good deeds in one's last birth—it is the family. And our scriptures also say:

a scoundrel is never born in a good family.
as glass is never found in a Padmarag mine.

Well, Tarlankar, my friend, you know the reference—ha! ha! ha! [*Takes snuff*].

SECOND TEACHER. We have been invited by Arabinda Babu of Saugandhya village; we shall honour Golak Chandra by accepting his hospitality for tonight.

NABINMADHAB. I am gratified Sirs, come this way.

All leave

ACT III

SCENE 1

Planter's office, Begunberey. Enter Gopinath Das and peon

GOPINATH. I know, I know, I'm never told anything. You come to me only when you're not satisfied with your share of the booty.

PEON. But tell me, Sir, can one digest all that muck by oneself? So, I told him, if you must eat, share it with our Mr Manager. Do you know what he said? 'Ha! you and your manager! Has he any power? He isn't that *kaibarta*'s lad, after all. That chap could call the tune and make those *sahebs* dance like monkeys.'

GOPINATH. Alright, go now. I'll show him what power this *kayastha*'s lad can wield!

Peon leaves

The swine has the junior *saheb* behind him, that's why all this big talk! It's fun to work if your brother-in-law is your employer, but that sort of talk, I've to say, makes the senior *saheb* furious. Yet the swine doesn't like me at all; for every little thing he threatens me with *Shyamchand*. Just the other day, he kicked me; thank God, he had his socks on. But recently, there's been some change for the better. Since Golak Basu was summoned by the court, he has softened a bit towards me. Only if you ruin your own people, are you 'clever' with these *sahebs*: 'You become a doctor when you first kill at least a hundred!' [*Seeing Wood at a distance.*] There he's! Let me first tell him about the Basus—that should soften him a bit.

Enter Wood

Sir, at last Nabin Basu's in tears; never before had that fellow been taught such a lesson. We had taken away the rascal's orchard; we had transferred his freehold land to Gadai Pode; we had practically stopped his cultivation. His barns are all empty. Twice we had dragged the swine to a criminal court. But he didn't bend. This time he's flat on his back.

WOOD. Bastard could do nothing in Shyamnagar.

GOPINATH. Sir, that clerk had gone to him, but it appears he told him, 'I'm not myself, all my strength is gone seeing Father in tears, they've put me in a fix'. Seeing him so crushed, seven or eight families of peasants have fled from Shyamnagar; the rest have begun to obey your Honour's orders.

WOOD. You good manager, your plan good.

GOPINATH. I always knew Golak Basu was a timid man; if we could drag him to court, he'd lose his mind. And the way Nabin Basu worships his father, he too will have learnt a lesson; that's why I suggested that the old man should be named as accused, and your new idea is also very good, Sir. Ploughing up the sides of his pond has stung him to the quick.

WOOD. Two birds killed one stone; ten *bigha* indigo and the sister-fucker miserable. That day swine very upset, said, have to leave house if indigo round pond, I said house land best for indigo!

GOPINATH. Hearing that, the swine has sued us.

WOOD. Case nothing—this magistrate good man—five years civil suit, case not end. Magistrate my good friend. See, our evidence accept. New law use, four rascals jailed. New law much better—*Shyamchand's* big brother.

GOPINATH. My Lord, Nabin Basu's helping the families of those four peasants. He has lent his own ploughs, cattle, and men to plough their land that they might not lose their crops.

WOOD. Bastard, not take advance, says no cattle, no ploughs for indigo, rascal very wicked, learnt good lesson. *Dewan*, good work, job continue, excellent.

GOPINATH. O personification of justice, it's all your kindness. I've plans to increase the amount of advance year by year, but it's not one man's job, one must have reliable overseers and peons. But can any work progress if we have a man who, just for two *rupees*, puts your worship to a loss of three *bigha* land?

WOOD. I understand, overseer swine done mischief.

GOPINATH. Sir, Chandra Goldar has recently come to the village, and hasn't yet signed for advance, our overseer goes and flings in his courtyard just one *rupee* as advance. The fellow cried, and begged him to take the *rupee* back and then followed him, pleading all the way up to Rathtala. There they met Nilkantha Babu, a lawyer, fresh from college.

WOOD. Yes, yes, I know, that bloody swine writes newspaper against me.

GOPINATH. His paper's nothing compared to yours, my Lord. You can't compare a tiny pitcher of cold water to a massive barrel! But Sir, your Worship had to spend a fortune to buy up that paper. Well, time's always a factor. They say there are times when one's relatives become strangers, and even a lame donkey fetches as much as a good horse.

WOOD. What Nilkantha did?

GOPINATH. He rebuked overseer soundly, he felt so ashamed that he went back to Goldar's house, and returned with two extra *rupees* and the one *rupee* he had given as advance. This Chandra Goldar's a devil, he could easily have sown indigo on three or four *bighas* of his land, but after all, is this the way a servant should go about his work? Only if I could work both as overseer and manager, instances of such base ingratitude would stop.

WOOD. Straight ingratitude—terrible wickedness.

GOPINATH. My lord, incarnation of justice, please forgive my boldness—this overseer had taken his sister to the junior *saheb*'s bedroom.

WOOD. Yes, yes, know, know—this bastard and that candy seller Padi corrupt junior *saheb*, shall teach swine lesson. Send that sister-fucker my sitting room.

Wood leaves

GOPINATH. Now see who can make the monkey dance better! Ha! ha! Two knaves—a sly *kayastha* and a sly crow! Now you'll taste a *kayastha*'s blow, your brother-in-law grandpa is beaten hollow!

SCENE 2

Nabinmadhab, Sairindhri sitting in their bedroom

SAIRINDHRI. Dearest, are my ornaments more important than my father-in-law? I know why you've been going around day after day; you don't eat, you don't sleep, your eyes are heavy with tears, you've lost that old cheerfulness and you've been getting these awful headaches—won't you let me give you these paltry ornaments?

NABIN. I know dear, you can give them easily, but it's not easy for me to take them. A husband goes into a lot of trouble to see that his wife can wear some nice ornaments—he doesn't mind facing a tiger, or living in a forest; he's ready to climb a mountain, dive into the sea, even swim across a turbulent river—he takes a lot of trouble to put a few pieces of jewellery on his wife. Am I so worthless that I'll take away her things? My love, let's wait today. Maybe I'll have to take these, if I can't get the money together.

SAIRINDHRI. We're passing through bad days dear; who'll trust us to lend us 500 *rupees* now? I beg you again—please mortgage mine and Saralata's jewellery and arrange for the money. *Chhoto bou*, my golden lotus, is miserable seeing you like this.

NABIN. Oh God, darling what are you telling me! Every word's piercing my heart like a burning arrow. Our *bouma* is just a girl, she'd love to have nice clothes, fine ornaments. Afterall, how can she understand what we've been passing through! Bipin cries, when, in fun you take away his neckchain; she too will cry if we take away her ornaments. God, God, I've become such an imbecile, such a cruel robber! To rob a little girl! Never—not as long as I'm alive. Even the cruel, inhuman planter won't do this. Please, don't ever again mention it dear.

SAIRINDHRI. Dearest, only I know and He from whom nothing is hidden, knows, how much it pained me to utter those words. You're right, every word's a fiery arrow; it pierced my heart too, burned my tongue, split my lips, before entering your heart. It was sheer misery for me even to suggest that we should take *Chhoto bou*'s ornaments. But to see you running round like mad, father-in-law in tears, mother-in-law sighing in grief, *Chhoto bou* utterly depressed, friends and relatives cheerless, even our tenants so disconsolate—can it even occur to me to have fun, to be merry? Only if we could get out of this trouble! And dear, it would hurt me to take away *chhoto bou*'s jewellery as much as it would to take away Bipin's. But it would be cruelty on my part, if I took Bipin's before I took her's; she'll be deeply hurt. She'll think I'm treating her as an outsider. Though I'm her sister-in-law, I'm like a mother to her; can I ever hurt her feelings?

NABIN. Dearest, you're wonderful; no two women are so straight, so guileless; ours was such a happy family—what was I, and what am I today! From our property our profit used to be seven-hundred *rupees* a year, fifteen barns bursting with rice. The orchard alone covered fifteen *bighas*, there were twenty ploughs, fifty labourers. How festive used to be the Durga Puja days! The house filled with guests, *brahmins* fed, food distributed to the poor! There were feasts for relatives and friends, *vaishnavs* sang, and the folk theatre brought such fun and merriment. It was a pleasure to spend money and to give in charity, even a hundred *rupees*, to the deserving. After all that wealth, I've been reduced to such a state today, that I've to sell the ornaments of my wife and sister-in-law. What an irony of fate! But why should I grieve? It was God who gave, and it's He who has taken it away.

SAIRINDHRI. There's such pain in my heart, dearest, when I see you sorrowing. It's my fate that you should suffer so much. [*She weeps.*] Please don't say no. [*She takes off her armlet.*]

NABIN. I too feel miserable if I see you in tears. [*He wipes away her tears.*] Don't cry, darling, don't cry. [*He holds her hand.*] Give me one more day, put it away, please.

SAIRINDHRI. I don't see any other way out—please listen to me. If our fortunes change you'll buy me, dearest, lots of ornaments.... [*Aside, some one sneezes.*] You see, what I said is going to come true.... Aduri's coming.

Enter Aduri with two letters

ADURI. I don't know from where these letters have come. Mistress asked me to give them to you.

Aduri leaves

NABIN. [*He opens one letter.*] I'll know now if I'll have to take your jewellery....

SAIRINDHRI. Please read it aloud.

NABIN. [*Reads the letter.*]

> This letter brings you my blessings.
> If I could lend you the money, it would be merely repaying an old debt. My mother passed away yesterday and I have to perform her last rites immediately. I had written to you about this yesterday. I could not sell the tobacco.
>
> Yours,
> Sri Ghanashyam Mukhopadhyaya.

Such bad luck! His mother's funeral isn't going to help me! Let me see what bombshell you've brought! [*He opens the second letter.*]

SAIRINDHRI. Hope turning into disappointment is a big blow! Don't read that one, dear.

NABIN. [*Reads the second letter.*]

> *Namaskar* from your dependent, Shri Gokulkrishna Palit.
> In all humility I beg to state, Sir, your good is my good, your letter has brought me all the information. I have been able to collect three-hundred *rupees* and shall be there tomorrow with the money. Next month I shall pay back another hundred. I will never forget how you helped me, and would like to pay some interest also.

SAIRINDHRI. Maybe, God has at last heard our prayers. I must go and give the good news to *chhoto bou* [*She leaves.*]

NABIN. Such a simple girl, so easily pleased. This is just a piece of straw in this raging flood—with this money let me take father to Indrabad now—later what has to happen will happen. I have one-hundred and fifty *rupees* in hand; if only we could put off selling the tobacco one more month, we would have made five-hundred, but I am afraid I'll have to sell it for three-hundred-and fifty. We shall have to pay a lot to the court clerks; the journey also will cost something, and if this false case ends in a sentence, disaster will strike this land. What a cruel law they've passed! But why blame the law or the lawmakers—if only the administrators of the law had been impartial, there wouldn't have been so much suffering! Such a pity! How many innocent people are in jail today, crying their hearts out; the pitiful sight of their families is heart rending. Empty pots on the ovens, grains parching in the yards, cattle tethered inside the cowsheds, fields lying half-tilled, most of them not even sown. The rice fields are full of weeds, how will the year pass? One hears only the cries of women and children! There are a few magistrates who are just, and in their hands this law has not become a death warrant; if only all of them had been like the one in Amarnagar! Could then the ripening crops of so many peasants have been destroyed—their corn-filled fields devastated by locusts? And would I myself have been in this predicament! O Lieutenant-Governor, if only you had thought of appointing good officers when you passed this law, there wouldn't have been such a disaster; only if there had been a Section, which would sentence the complainant to jail, when his complaints were proved to be false, the Amarnagar prison today would have burst with indigo planters, and they wouldn't have been so powerful. Though our magistrate has been transferred, our case is going to be heard here. I'm afraid that'll be our end.

Enter Sabitri

SABITRI. Nabin, suppose we sell off all our ploughs, do we still have to take advance? Listen, please, sell off all our cattle and ploughs and start some business. Our income will be more than enough for us. This is becoming unbearable.

NABIN. Mother, that's what I too would like to do. I'm waiting for Bindu to get a job, but just now if we stop all cultivation it'll be difficult to run the house; that's why, inspite of such ordeal, I've kept some of the ploughs.

SABITRI. How will you go on this long journey with this headache? Oh God, what a punishment, this indigo! [*She massages his head.*]

Enter Rebati

REBATI. Oh *ma, ma,* what shall I do, where shall I go, why did I ever bring her here, when she was no longer mine and belonged to someone else and when I couldn't even give her protection? *Baro babu*! Save me, my heart is breaking—bring back to me my Kshetramani—bring me back my golden doll.

SABITRI. What's happened? What's the matter?

REBATI. This afternoon Kshetra had gone with Pencho's *ma* to Dasdighi to fetch water; on her way back through the garden four musclemen came and kidnapped my little girl. That harlot Padi pointed her out to those men and fled. I was looking forward so much to having a little celebration for her sixth month ceremony. Why did I bring her back now that she's somebody else's—why did I *baro babu*, why did I?

SABITRI. This is terrible! These monsters can do anything these days—they're snatching away people's land, robbing them of their rice, their cattle, wielding *lathis*. They're forcing them to sow indigo—though in tears, utterly miserable, they're sowing. And now they've started destroying their caste!

REBATI. *Ma*, though half-starved, we're sowing their indigo. All the *bighas* they mark out, we're sowing on them all. Ray, poor lad, ploughs the land and weeps. When he comes back from the field and hears this, he'll go mad.

NABIN. Where is Sadhu?

REBATI. He's sitting outside, crying.

NABIN. Father's Bhim[1] of Swarpur is still alive. Yet women are being carried off. To a woman of good family her chastity is her lodestar. I must rush—let me find out that Dushashan. Her chastity, white like a lotus, no indigo-frog will be allowed to pollute!

SABITRI. Chastity is gold—a gift of God; a beggar woman's a queen, if she has this treasure. If you can rescue, untarnished this gem in all its purity, from the grip of the indigo monkey, I'll feel proud to have given birth to such a son. This sort of oppression is unheard of. Come Ghosh *bou*, let us go out.

All leave

SCENE 3

Rogue sitting in his bedroom; enter Padi and Kshetramani

KSHETRA. Auntie, auntie don't ask me to do that—I'll give my life, but not my *dharma*. Cut me to pieces, burn me to ashes, drown me, bury me alive, I won't let a man, who's not my husband, touch me! How'd my husband feel?

PADI. Now, don't be silly. Where's your husband, and where're you? Nobody'll ever get to hear of it. I myself'll take you back to your mother this very night.

KSHETRA. Maybe my husband won't get to know about this, but there's God above, who gets to know every thing—I shan't be able to cheat Him. A fire will burn inside me—like a kiln-fire. Every time my husband loves me as his chaste and good wife my whole being will be on fire; whether he knows it or not, I'll never take a paramour.

ROGUE. Padda, why aren't you bringing her to the bed?

PADI. Come dear, go to *saheb*; tell him whatever you want to say. To tell me anything is to cry in the wilderness.

ROGUE. Well, an appeal to me is like scattering pearls before swine. Ha! ha! ha! We're indigo planters. Yama's our friend, we stand and burn down so many villages; so many mothers while giving breast to their children burn and die before our eyes! Does that make us sad? Can a factory run if we feel sentimental? But we're not really bad people as such, it's our work with indigo which brings out the evil in us. First time we killed a man we felt bad, but now my *ramkanth* beats ten women unconscious—and next moment we enjoy a good dinner! But I love women, and factory work is good for that kind of job—all mix in the sea. You've no strength Padda? Pull her, drag her to me.

PADI. Kshetramani, my good girl, come, sit on the bed. *Saheb* will give you a lovely gown, which *bibis* wear.

KSHETRA. To hell with a *bibi*'s gown! I'd rather wear a strip of sacking, than a *bibi*'s gown. Auntie, auntie I'm so awfully thirsty, please take me home, a drink of water would cool me a little. God, God, mother must have hanged herself by now, and father too must have chopped his head off with his axe, and my uncle must be running around like a wild bull! They've nobody but me, I'm their only child. Let me go, take me home, let me touch your feet Padi auntie—oh mother, I'm dying of thirst.

ROGUE. There's water in the pitcher.

KSHETRA. I'm a Hindu woman, I don't touch *saheb*'s water—those musclemen have touched me—first I'll bathe, and only then enter my house.

PADI. [*Aside*] I've lost both my *dharma* and my caste. [*Aloud*] Well dear, what can I do? Once you're in their clutches, it's difficult to get out. *Chhoto saheb*, let Kshetramani go home today, she'll come some other day.

ROGUE. Yes, yes and you've fun with me? Right? Get out, get out, I'll soften her. If I can't I'll send her back with you. You damned whore, now I know, I'm sure it was you who didn't let her come, so I had to send musclemen to fetch the daughter of those good people. Do I ever easily ask our musclemen to do such jobs? You scoundrel! Whore of a candy-seller Padi!

PADI. Send for that Kali, she has suddenly become your pet, it's very clear now.

KSHETRA. Oh auntie, auntie, don't go, don't go away.

Padi leaves

You've left me alone in this cobra's pit! I'm so frightened, I'm trembling, my whole body's reeling in fear, my mouth is dry, great thirst is killing me.

ROGUE. Dear, dear. [*Pulling her by both her hands.*] Come, come my angel.

KSHETRA. Oh *saheb*, oh *saheb*, you're like my father, please let me go. Send me back with Padi auntie, the night's dark, I shan't know my way. [*Rogue continues to pull her.*] Oh *saheb, saheb*, you're like my father, I lose my caste if you touch me—let me go, *saheb*, let me go.

ROGUE. I'll be your brat's daddy—don't try sweet talk—come to bed—else I'll kick your belly.

KSHETRA. *Saheb, saheb*, my child will die! I beg you *saheb*, my baby'll die, I'm pregnant.

ROGUE. [*Pulling at her* sari.] Must strip you, then you won't be squeamish.

KSHETRA. Don't strip me *saheb*; I'm your mother, you're my son. *Saheb saheb*! Leave my *sari* alone. [*She digs her nails into his hands.*]

ROGUE. You infernal bitch! [*He takes the whip down.*] This'll teach you a lesson.

KSHETRA. Yes, kill me! I shan't say a word. Cut me down with your sword, I'll go straight to heaven. Don't touch me—you whore's son! You son of a pig! May death visit your house! If you touch me again, I'll bite your hand to bits! Have you no mother, no sister... go and take off their clothes! Why're you standing like that? You harlot's brother, kill me, kill me! I can't bear this any more.

ROGUE. Shut up you bitch, mouthing such big words! [*He pulls her by her hair and punches her in the abdomen.*]

KSHETRA. Oh father, oh mother, where are you? This is the end of your Kshetra. [*She shivers.*]

Enter Nabinmadhab and Torap, breaking open the window shutters

NABIN. [*Releasing Kshetramani's hair from Rogue's clutches.*] You planters are so low, such monsters—hasn't your Christianity taught you to control your lust? Is this an example of Christian kindness, humility and morality? My poor, poor child! What a brute you've been to a helpless, pregnant girl!

TORAP. Look at the swine—standing like a wooden statue—bastard has lost his tongue, he's not the one to listen to scriptures! *Baro babu*—as the dog, so the rod and I have it—that swine's huge jaw's just waiting for a huge slap. [*He siezes him by his throat, and hits him.*] If you shout, I'll send you straight to Yama's house. [*Presses his cheeks hard.*] We've a saying, 'A thief has five days, a good man one'. For five days you've fed us on blows—today you eat this [*twists his ears*], you swine!

NABIN. There's nothing to be afraid of, get dressed. [*Kshetramani wears her* sari.] Torap, you keep the scoundrel gagged, I'll carry this girl and run home. Once I've gone beyond the tribal village, you let him off and then run for your life. It's difficult to go along the river bank—thorns have scratched me all over—the tribals must be sleeping now, even if they're awake, they won't harm us, once they hear this. But you better come to our house, I'd like to hear how you fled from Indrabad, and where you're staying now.

TORAP. Tonight I'll swim across the river. How can I tell you, *baro babu*, what I suffered—somehow I managed to break open the stabledoor of that swine of a lawyer—and ran straight to Basantababu's estate. Then at night, I went back and brought my wife and children to the new home. These bastards have forced me to leave my village. Can we ever again plough our own fields for our own food? The demand for indigo has ruined us all—and then on top of all this—they tell us to betray our benefactors! You swine, why aren't you now using those words gad, mad and what not and kicking us with those boots? [*He kicks Rogue.*]

NABIN. Torap, why beat him? If they are brutes, we don't have to be. I'll be off.

Nabin leaves with Kshetra

TORAP. You rascals, you want to ruin such men; talk it over with that big dad of yours. Come to some agreement and then carry on. How long can you go on crushing people? If they all run away, what can you do? After all, once a man's dead, nothing can be done about it! You swine, once the peasants leave the villages, your *kuthi* will be your grave. Pay back *baro babu* his last year's dues and agree to take the indigo they'd like to sow for you. It's because of you they're facing so much trouble. Merely forcing advances down one's throat doesn't help—one must till the earth *chhoto saheb*! *Salam*—I'm going.

He knocks him down flat and runs away

ROGUE. By Jove! Beaten to jelly!

SCENE 4

Veranda in Golak Basu's house: Enter Sabitri

SABITRI. [*Sighing*] Oh cruel *hakim*, why didn't you summon me too? I could then have gone to the district town with my husband and son! That would've been far better than living in this graveyard. Ah my misfortune, my *karta*'s.[2] A home-bird. Even when invited for dinner in another village, he doesn't accept; today that man has been summoned by a criminal court as an accused—he might be sent to jail—was it his destiny to suffer! Bhagabati Devi,[3] Mother, is this your wish? Ah me, he can't sleep unless the room is spacious, he eats only the sun-dried rice, and that too cooked only by *baro bouma*; when the news came he began to beat his breast so hard that blood oozed out, and his eyes swelled with tears. And those parting words still ring in my ears—'Wife, this is my last journey—to the Ganga[4]. [*She weeps.*] Nabin tells me, 'Mother, pray to Bhagabati, I'm sure to win, and shall bring him back home.' My darling's fair face has become dark with fatigue, and the trouble he is having to raise money—this constant running around's causing him giddiness, but lest I give him the ornaments of my daughters-in-law to sell, he tells me, to give me courage, 'Mother, we're not that short of money—afterall, how much do we need for the case?' That he had to mortgage my ornaments during the last case about our land and property, has made him so sad. And he tells me, 'The day I get some money, the first thing I'll do is to get those ornaments back'. His words are brave, but his eyes are full of tears—my darling left weeping, in this blazing sun, my Nabin has to go to Indrabad, and I'm sitting here in comfort and I'm his mother!

Enter Sairindhri

SAIRINDHRI. *Thakrun*, it's very late, please go and have your bath. Misfortune has hit us, why else would such things happen to us!

SABITRI. [*Weeping*] *Ma, bouma,* before my Nabin comes back, I shall not eat nor drink. Who's going to feed my child?

SAIRINDHRI. Brother-in-law has a room there, and also a *brahmin* cook. There won't be any trouble, you go and have your bath.

Enter Saralata with a pot of oil

Chhoto bou, rub a little oil on *thakrun*; when she has bathed, bring her to the kitchen, I'll get the food ready.

Sairindhri leaves. Saralata massages Sabitri with oil

SABITRI. My little parrot's silent today, my darling's so quiet! The lovely face is like a drooping flower, and ah! it's so long we haven't seen Bindumadhab. His college is closed, and we were all looking forward to my darling's home-coming so eagerly. And now this! [*Fondling her.*] My child you look so tired, you haven't eaten anything I think. I'm in such agony, that I don't even see if my girls have eaten or not. I shall bathe myself, you go and eat something, I shan't be long.

Both leave

ACT IV

SCENE 1

Indrabad, the Magistrate's court: Wood, Rogue, Magistrate, Court officials sitting; Golak Chandra, Nabinmadhab, Bindumadhab, Defence and Prosecution Lawyers, Court supervisor, peons, orderlies, peasants, etc., standing

DEFENCE LAWYER. I pray that the petition of the accused may be allowed. [*He hands over the petition to the head clerk.*]

MAGISTRATE. All right, read. [*He consults Wood, and laughs.*]

HEAD CLERK. [*To Defence Lawyer.*] A veritable Ramayana you have written! Can one read a petition unless it is short! [*He turns over the pages of the petition.*]

MAGISTRATE. [*Stops talking to Wood, and controls his laughter.*] Read it through.

HEAD CLERK. [*Reads aloud.*] The evidence for the prosecution was recorded in the absence of the accused and his lawyer; it is prayed that witnesses for prosecution be produced in the court for being cross-examined.

PROSECUTION LAWYER. My lord, lawyers are well known for swindling, cheating and telling lies; they make false statements, even while on oath, but they do much worse things—they desert their own wives and spend their time in their heaven, with prostitutes. Actually *zamindars* detest these fellows, but to get their work done, they call them and sit together. My lord, their profession is to cheat. But there is a difference—lawyers engaged by planters are compelled to base their cases only and exclusively on truth. This is because planters are Christians; falsehood is deemed to be one of the worst sins in the Christian religion. Plunder, adultery, murder are acts hated in this religion; not only can a Christian not act dishonestly, he cannot even think dishonestly; if he does, he burns in hell-fire. Christianity demands compassion, forgiveness, humility, and philanthropy. The planters, who have imbibed this great truth and are religious, can never give false evidence. And, My lord, we lawyers, who are in the pay of such men, have followed their examples and reformed our habits and have become better. Even if we wish to train a witness to give false evidence, we dare not, for these truth-loving *sahebs* punish us severely, on the merest suspicion of falsehood. This witness for the defence, who used to be an *amin* of the planters is an example. It has been recorded that this man had cheated a peasant of his advance, and was dismissed. Moved by the poor peasant's tears the kind *saheb* had thrashed this fellow in anger.

WOOD. Extreme provocation, extreme provocation.

PROSECUTION LAWYER. My lord, the court has questioned our witnesses. Had they been tutored, they would have been found out by your Honour. According to lawmakers. 'The judge is the advocate of the accused'. So your Honour has already questioned witnesses on behalf of the accused. If therefore, those witnesses are produced in the court again, it will be of little benefit to the accused, and the witnesses will be put to great inconvenience. My lord, these men are poor cultivators, they themselves plough their fields so that their families can survive—one day's loss of work in the field, means total loss of their crops. To prevent any loss, they do not come home for lunch, their women carry their food to the field and feed them there. Just a day's absence from his field, ruins a cultivator. If these men are summoned from their distant fields, they will lose the produce of their one year's labour. But I leave it to your Honour to decide.

MAGISTRATE. No justification whatsoever. [*Consults Wood.*] Not necessary.

DEFENCE LAWYER. Your Honour, in no village, does a peasant voluntarily accept an indigo planter's advance. Accompanied by his overseer and employees, the planter or his manager, rides to the peasants' fields, marks out their best lands and orders them to sow indigo there. These peasants are then dragged by his men to his factory, and compelled to take advance. When they go home grieving, there is such lamentation that one might think someone in the family has died. Even after the peasant has supplied indigo over and above what he owed the planter, the register continues to show him in debt. Once they take advance, they suffer at least for seven generations. Only they know and God above—the Protector of the poor—knows what misery it is for them to have to sow indigo. When they get together, the only topic of conversation among them is the forced advance each had to sign for, and they discuss different means of escape from the clutches of the planter. They don't have to be advised by others—they know—and are running amuck like wounded dogs, mad with pain. Under these circumstances, it is amazing and palpably false, that so many peasants could have come before the court and deposed that they were always eager to grow indigo, but could not, because my client had advised them not to sow and threatened them if they did. Your Honour, if they are produced before the court once again, I shall ask them only one or two questions and prove that they have given false evidence. I concede that my client's son, Nabinmadhab Basu, does his best to protect these helpless, miserable peasants, from these beasts of night, the terrible indigo planters, and many times he has been able to save them also from the tyranny of Wood *saheb*. The records of the case of arson in Palaspur will conclusively prove this. But my client, Golak Chandra Basu, is a very gentle and timid person. He fears planters more than tigers, he has never been in any trouble, and has never harmed any one in his life; nor does he dare go out of his way to help others when they are in trouble. Your Honour, every one knows Golak Chandra Basu in this district as a very fine man and respects him highly; this can be verified from your own officials.

GOLAK CHANDRA. Your Honour, I have yet to be paid for last year's supply of indigo, but anxious to avoid coming to the court, I even agreed to accept an advance for another sixty *bighas*. My elder son said, 'Father, we have other sources of income. Our loss for a year or two because of indigo will prevent us from meeting only our religious and social commitments. But we shall not starve. But what will happen to those whose only means of subsistence is their plough? If we agree to sow indigo at this rate, they'll all have to agree.' These were wise words, and I could not say 'no' and asked him to go to *saheb* and beg him, entreat him to agree for fifty *bighas*. But the *saheb* said nothing—neither 'yes' nor 'no'. Instead, he quietly brought this false case against me, to send me to jail in my old age. I realize that the only way to safety is to be on the right side of the *sahebs*. The *sahebs* own this country now, the magistrate is their brother; can we afford to go against them? Please release me, I swear even if I cannot sow indigo for want of ploughs and cattle, I shall pay the *saheb* one-hundred *rupees* every year, in lieu of indigo. I am not the man to instigate the peasants—do I ever come in contact with them?

DEFENCE LAWYER. Your Honour, of the four witnesses for the prosecution, one is a day labourer—his family never owned a plough. He has neither land, nor cash, nor cattle; not even a cowshed; an enquiry on the spot will prove this. Kanai Tarafdar belongs to another village altogether; he has never met my client, nor will he be able to identify him. For these reasons I pray, that Your Honour may be pleased to order their production before the court. Law demands that the accused be given every facility before the judgment is delivered. If Your Honour grants this prayer, we shall be forever grateful.

PROSECUTION LAWYER. My lord....

MAGISTRATE. [*Writing*] Go on, I'm not writing with my ear.

PROSECUTION LAWYER. Your Honour, if it had not been that these poor men will be put to great hardship, I too would have asked them to be produced. I could then have questioned them and established the guilt of the accused more firmly. Your Honour, all over the country people know about Golak Basu's wickedness—he bites the hand that feeds him. Crossing the mighty ocean the planters have come to this country; discovering the secret treasure they have brought prosperity to the land, increased the revenue of the Government, and then improved their own fortune. The man who acts against such noble people, where else should he be sent but to prison?

MAGISTRATE. [*Writes the address on the letter.*] Peon!

PEON. [*Approaching*] My Lord.

MAGISTRATE. [*Consults Wood.*] Take this to Mrs Wood. Tell the cook that the guests won't come today.

Exit the peon

HEAD CLERK. Your Honour, what order should be passed?

MAGISTRATE. The petition should be filed.

HEAD CLERK. [*Writes down.*] It is ordered that the petition be filed. [*Magistrate signs the order.*] Your Honour has not signed the order on the statement of the accused.

MAGISTRATE. Read it.

HEAD CLERK. [*He reads.*] 'It is ordered that the accused shall furnish a security of two-hundred *rupees* with two sureties for like amount. Issue summons to the defence witnesses.' [*Magistrate signs.*]

MAGISTRATE. The case of the dacoity at Mirgaon is adjourned till tomorrow.

The Magistrate, Wood, Rogue and orderlies leave

HEAD CLERK. [*To court supervisor.*] See that the security bonds are properly executed.

Head clerk, prosecution lawyer and peasants leave

COURT SUPERVISOR. [*To the defence lawyer.*] It's almost evening. I don't see how the bonds can be executed today—besides, I'm a little busy.

DEFENCE LAWYER. The family's well known—but they've nothing. [*He talks to the court supervisor.*] This money has to come from the sale of ornaments.

COURT SUPERVISOR. You see, I don't possess either a *taluk* or arable land, nor have I any business. This is my only means of livelihood. And it's only because of you that I agreed to a mere hundred *rupees*. But you must come to my house. See that the manager doesn't get scent of it—their propitiation will have to be done separately.

All leave

SCENE 2

Indrabad, Bindumadhab's rented house—Nabinmadhab, Bindumadhab and Sadhu Charan sitting together

NABINMADHAB. I'm afraid I've to go home now. When mother hears this news it'll kill her. Well Bindu, what can I tell you, see that father doesn't have to face any kind of trouble. I've decided to quit the village. I shall sell off every thing we have and send you the money. Pay whatever they demand.

BINDUMADHAB. The jailor himself isn't after money, but he's afraid of the magistrate, and wouldn't let the *brahmin* cook into the jail.

NABINMADHAB. All the same, pay him, fall at his feet. It's so awful—at this age to be without food for three days! I begged him, tried to persuade him—but all he said was, 'Nabin, I will decide to eat or not only after three days are over. No food will enter this cursed mouth before that.'

BINDUMADHAB. I really don't know how to make him eat a few grains of rice. The moment that planter's slave, that stupid magistrate, passed the cruel order sentencing him to jail, he put his hands up to his eyes, and they're still there—wet with tears—and he hasn't moved from the place where we had first made him sit. Silent, frail, motionless, he's like a dead bird inside this cage of the jail. It's the fourth day today, somehow I must see he eats something. You go home, I shall write to you every day.

NABINMADHAB. Oh God! You are making him suffer much! Bindu, if they permitted you to keep him company inside the jail day and night, I could return home with a little less worry.

SADHU CHARAN. Let me steal something, you drag me to the court as a thief, I shall admit my guilt, then they'll put me in jail, there I shall serve Master as his servant.

NABINMADHAB. Sadhu, you're really a *sadhu*. I know how upset you are about Kshetramani's grave illness. I must take you back as quickly as I can.

SADHU CHARAN. [*Sighing*] *Baro babu*, shall I see my child again? She's all I have.

BINDUMADHAB. Give her the medicine I got, she'll certainly feel better. The doctor prescribed this after I explained the whole case to him.

Enter the Deputy Inspector

DEPUTY INSPECTOR. Bindu *babu*, the commissioner *saheb* has made a special recommendation that your father be released.

BINDUMADHAB. I'm quite sure that the Lieutenant-Governor will order his release.

NABINMADHAB. How long do you think it'll take for the order to reach here?

BINDUMADHAB. It shouldn't take more than a fortnight.

DEPUTY INSPECTOR. The magistrate of Amarnagar had sentenced a lawyer to six months under this Act. He had to spend only sixteen days in jail.

NABINMADHAB. Will there be such a day for us—when the Lieutenant-Governor will see our point and set aside this terrible order of the hostile magistrate?

BINDUMADHAB. God is all merciful. He will certainly do that. You have to travel a long distance, you'd better start now.

Nabinmadhab, Bindumadhab and Sadhu Charan leave

DEPUTY INSPECTOR. Such a pity! These two brothers are half dead with grief and worry, only the release order from the Lieutenant-Governor will put back some life into them. Nabin *babu* is a very courageous man—generous, unselfish, always eager to help others; he has done so much to spread education—a real patriot, but all these fine qualities of his, like buds in a storm, are being crushed by the cruel planters.

Enter a pandit of the college

Come in, come in, Sir.

PANDIT. I couldn't come even once during these terrible days

Bindumadhab has been going through; my body heat is constitutionally a little high,[1] and I cannot stand the sun. The heat of the months of *chaitra* and *baishakh* make me almost mad. For the last few days I have been having dreadful headaches.

DEPUTY INSPECTOR. I think Vishnu oil will be good for you. We prepared it for Bishnu *babu*. I shall send you some tomorrow.

PANDIT. Thank you very much. Even normal, healthy people go mad teaching these boys, and now of course I am in a very bad state of health.

DEPUTY INSPECTOR. I haven't seen the senior *pandit* for some days now!

PANDIT. Well, he is now planning to give up this dog's life. His son—a brilliant lad—has started earning, he can now live like a king. Besides it does not look nice to go to college every day with a bull's halter tied round one's neck[2] and he is no longer young either.

Enter Bindumadhab

BINDUMADHAB. Oh Sir, you have come.

PANDIT. That devil has been so unfair. You haven't heard perhaps—he had spent all the ten days during Christmas as that planter's guest. And then to expect that he will be just to the people! It is like a Muslim judge passing judgement on a religious festival of the Hindus!

BINDUMADHAB. It's God's will.

PANDIT. Who was your lawyer?

BINDUMADHAB. Pradhan Mallick.

PANDIT. He of all men! Anyone else would have been better—but then all Gods are the same. Run around the village to find one honest chap!

BINDUMADHAB. We hear the commissioner *saheb* has recommended father's release to the Government.

PANDIT. Not much difference—cinders and ashes—who is good and who is bad! Like magistrate, like commissioner.

BINDUMADHAB. You are saying this, Sir, because you do not know the commissioner well. He's absolutely impartial and wants the natives to progress.

PANDIT. Let us hope, God willing, for everyone's good, that your father can be rescued from this mess. How is he in the jail?

BINDUMADHAB. He weeps all the time, and hasn't eaten a bite in the last three days. I'm going to him now and shall give him this good news to cheer him up a little.

Enter a peon

Aren't you the peon of the jail?

PEON. Sir, please come to the jail immediately. The guard wants you there.

BINDUMADHAB. Have you seen my father today?

PEON. Please come, I can't say anything.

BINDUMADHAB. I'm coming. [*To the* pandit.] Something awful must have happened, I must hurry.

The peon and Bindumadhab leave.

PANDIT. Come let us also follow them, something serious must have happened.

Deputy Inspector and pandit leave.

SCENE 3

Indrabad jail Golak Chandra's body is seen hanging from the roof, tied to a scarf. Police constable and jail warden are standing

CONSTABLE. Who has gone to call Bindumadhab *babu*?

JAIL WARDEN. Maniruddi has gone. Till the doctor comes we can't take down the body.

CONSTABLE. Isn't the magistrate *saheb* expected today?

JAIL WARDEN. No Sir, not for another four days. On Saturday there's going to be a champagne party in the planter's place at Sachiganj; then there'll be a dance. Wood *saheb*'s *bibi* dances only with our *saheb* and with nobody else—I used to see this when I was their orderly. *Bibi* is a very kind lady, one single letter from her made a poor chap like me a warden of the jail.

CONSTABLE. What a pity this! Bindu *babu* was feeling so wretched because his father hadn't eaten for three days—now when he sees this, it'll kill him. But every thing happens according to His will.

Enter Bindumadhab

BINDUMADHAB. Oh God, oh God, what's this, what's this! Has Father hanged himself! Just as I was coming with the news of this possible release! What unbearable agony! [*He embraces his father's body, and cries.*] Father's gone—he has left us forever. There is nobody now to feel proud of his Bindumadhab's knowledge of English, none to call his Nabinmadhab 'Swarpur-Brikodar', none to call *baro bou* 'my mother, only my mother' to tease Bipin and laugh with him like two old friends—all that's over. And what will happen to my mother when the news reaches her! When she hears this news she'll be as grief-stricken as the mother crane, whose companion was shot down by a cruel hunter, when they went looking for food to feed their offspring.

CONSTABLE. [*Leading Bindumadhab by the hand.*] Bindu *babu*, try to be calm. Ask for the doctor's permission and arrange for the cremation in the Amritaghat.

Enter Deputy Inspector and pandit

BINDUMADHAB. Constable, Sir, please leave me out; kindly fix up whatever's necessary, consulting *pandit mahashaya*, and the Deputy *babu*. I can't talk, let me hold my father's feet to my heart for the last time and be with him. [*He sits down, pressing his father's feet to his breast.*]

PANDIT. [*To the Deputy Inspector.*] I shall mind Bindumadhab; you please untie the cord, one shouldn't keep the body of such a godly man in this hell even a second longer.

CONSTABLE. Sir, please wait a little longer....

PANDIT. Oh, I suppose you are the keeper of this hell? How else could you be so inhuman?

CONSTABLE. Sir, you are a learned person, but you are rebuking me unfairly.

Enter the doctor

DOCTOR. Hallo Bindumadhab, God's will you know! Oh, *pandit mahashay* also is here! Bindumadhab isn't going to give up his studies, I hope.

PANDIT. He certainly should not.

BINDUMADHAB. We've lost all our property, and now Father's gone, we're left with nothing. [*He weeps.*] How can I continue my studies?

PANDIT. Planters have robbed them of every thing.

DOCTOR. Yes, I've heard from the *padris* about these planters and I myself also saw something the other day. I was returning from the Matanganagar factory, and while stopping at a village, saw two peasants pass by my palanquin on their way to the market; one was carrying milk. Wanting to buy some, I called him; then I saw one chap whispering something to the other, and then both just dropped the milk and ran for their lives, shouting 'Indigo Devil, Indigo Devil'! I told this to another peasant, and he said, 'They had fled because they were afraid of the advance, but I've already taken advance, so I shouldn't be put inside the lock up.' I realized then that I had been taken for a planter—so I gave the milk can to the man and left.

DEPUTY INSPECTOR. Yes, one day a *padri* was passing through a village which belonged to the concern of Bailey *saheb*. The moment the peasants saw him, they left the road and just bolted for their own homes, screaming, 'The indigo devil has come out, the indigo devil has come out'. Gradually however, the peasants were surprised to find the *padri* so benevolent and courteous and forgiving. The more the *padri* expressed his genuine sympathy for the suffering of the peasants, oppressed by the planters, the more their regard for him grew. Now they say, 'Though they're bamboos of the same clump, some are used for the frames of Durga's image and some for the sweepers' baskets of shit.'

PANDIT. Now let us remove the body.

DOCTOR. I'll have to examine it a little; please take it outside.

Bindumadhab and the Deputy Inspector take down the body—all leave.

ACT V

SCENE 1

In front of the Begunberey factory office: Enter Gopinath Das and a milkman

GOPINATH. How did you get all this news?

MILKMAN. We're neighbours—when we're short of salt we run there to get some, when we run out of oil we ask them for it, when the child cries, we go across to fetch him some treacle. For seven generations we've lived on the charity of the Basu's. Who else should know better what's going on there?

GOPINATH. Where was Bindumadhab married?

MILKMAN. Some village, west of Calcutta—I forget—that place where they want to put the sacred thread[1] on the *kayasthas*. Fancy, wanting more Brahmins! Already their number is legion and really impossible to feed. *Chhoto babu*'s father-in-law's a big man, even the Governor *saheb* has to take off his hat when they meet. Do such men ever give their daughters in marriage in a village? But he didn't bother about this tiller's village of ours—do you know why—because he saw how good *Chhoto babu* was at his studies. They say that city girls are rather coquettish, and you can't tell them from those *bazar* women—but you'd never find a girl as quiet and gentle as the daughter-in-law of the Basu's; Goma's mother goes to their house every day, but she never once saw her face in the last five years that she's been married, and even we saw her only once—the day she came here as a new bride. All of us had thought that city people follow the ways of the English, and bring up their daughters like the *bibis*.

GOPINATH. We hear she looks after her mother-in-law all the time.

MILKMAN. Manager, Sir, what shall I say—Goma's mother told me, and it's common knowledge in the village—that but for *chhoto bou, mathakrun* would have killed herself, the day the news of the hanging reached her. We had heard that city women turn their husbands into sheep, and starve their parents-in-law to death, but now that we've seen this *bou*, we realize that such talks are just rumours.

GOPINATH. Nabin Basu's mother also is very fond of this *bou*, we hear.

MILKMAN. Well, I don't know whom *mathakrun* doesn't love in this wide world. That woman is Annapurna[2] herself. But have you people left her with any rice, which she can give to others? The cursed indigo has eaten the old man and it won't be long before it eats up the old woman also....

GOPINATH. Shut up, you fool, if the *saheb* hears this he'll make you shit blood.

MILKMAN. What could I do? It's you who started goading me, ferretting out this poisonous stuff! Do you think I like to abuse that swine, sitting in his *kuthi*?

GOPINATH. I'm feeling a little unhappy. You see, by bringing that false case I've destroyed that noble man, and I'm pained to hear that Nabin has been suffering from such severe headaches, and Nabin's mother is so miserable.

MILKMAN. Crocodile tears![3]—Now Mr Manager, Sir, don't get angry—I'm a mad chap as you know—shall I get your hubble-bubble ready?

GOPINATH. You rascal, you're the true son of the Nanda's[4]—there's no limit to your teasing!

MILKMAN. The *sahebs* have started doing all sorts of things—they're the blacksmiths, and you're their swords, you strike as they order. May the swine's *kuthi* go under water! All will feel better after a nice dip there.

GOPINATH. You rascal are really a silly ass—I've heard enough! Be off! *Saheb* will be here any time now.

MILKMAN. Yes, I'm going, please keep my milk-account ready. We're going tomorrow for a bath to the Ganga, I'll need a *rupee*.

Milkman leaves

GOPINATH. Tomorrow, I think, the lightning will fall on his already ailing head. They'll sow indigo all round their pond—and nobody can stop it. The *sahebs* are a little wrong there—afterall they had agreed to sow indigo on fifty *bighas*, though they hadn't been paid their dues for last year. Even that didn't satisfy! All this trouble arose because of those paddy lands to the east—Nabin Basu ought to have handed those over—it's always wise to be on the right side of Ma Sitala.[5] But Nabin will bite—even if it's his last bite. [*Seeing the planter at a distance.*] There he's—the white angel, all dressed up in a blue suit. God knows if I too will be lodged for sometime with the old manager!

Enter Wood

WOOD. No man must know. A very big flare up Matanganagar *kuthi*—all *lathiyals* there—here ten Pode[6] spear men keep ready—I go, *chhoto saheb* go, and you go—that swine, mourning the cloth round neck, can't much do—also ill—how can get constable's help!

GOPINATH. Those bastards are in utter distress—it's not necessary to keep spearmen here. Among the Hindus suicide by hanging, and specially inside a jail is particularly heinous. It's a terrible punishment for them.

WOOD. You don't understand—rascal happy father died—he took advance because afraid father, now sister-fucker no fear, do what he like—swine brought bad name my estate. Tomorrow I arrest him, with Majumdar he live. Amarnagar Magistrate type magistrate come, bastard begin mischief.

GOPINATH. The course Majumdar's case has taken would have been much more serious, if Nabin Basu had not been in this trouble. Even now we can't be sure; we hear that the new magistrate who is coming supports peasants and camps out while inspecting villages. That might mean trouble and be a cause for fear also.

WOOD. You swine bothered me—always your 'fear, fear'. Indigo planter afraid, any work? You jackal, bastard, not like job, get out!

GOPINATH. My lord, I do feel a little frightened—when the last *dewan* went to jail, his son came for his pay which was in arrears for six months; you told him to send an application. After he did that, you ordered that no payment would be made till he produced written proof. So My lord, when servants go to jail, this is the justice they get!

WOOD. Oh! I not know? You swine, ungrateful son of a bitch! Salary nothing to you! If you not eaten indigo price money not this deadly Commission[7] set up! Not go sorry peasants crying, crying *padri sahebs*. You bastards ruined everything. If material short weight, I sell your house, get back money. Arrant coward! Hellish knave!

GOPINATH. Sir, we are butcher's dogs. We fill our stomachs with entrails. Your Honour, if you people had acquired the indigo as the money lenders acquire their rice from the debtor, the indigo estates wouldn't have earned such a bad reputation, and it wouldn't have been necessary to engage overseers and peons, and I too wouldn't have been rebuked by the people and called 'Gupe shit-eater, Gupe bastard'.

WOOD. You bastard are blind, no eye.

Enter a candidate for job

I saw own eyes. [*Pointing to his eyes.*] Money lenders go rice fields quarrel with peasants. Ask this man.

CANDIDATE. Your Honour, I can give many such instances. Peasants say that because of the kindness of the *sahebs* we are feeling safe from the clutches of money lenders.

GOPINATH. [*Aside to the candidate.*] My dear man, you're flattering for nothing, there's no vacancy. [*To Wood*]. True, money lenders do go to the paddy fields and argue and quarrel with their debtors, but if the real significance of their visits and quarrels had been properly understood, you wouldn't have compared *Shyamchand*'s crushing blow on starving peasants—as powerful as the weapon which had killed Queen Sumitra's son[8]—with the visits in the paddy fields of the large-hearted money lenders, well wishers of the peasants. We're very different from them.

WOOD. All right, explain, some reasons are—why bastards talk all things us only, but never money lenders.

GOPINATH. My lord, all the money they need for the whole year, they borrow from them; all the rice they need for their food, they take from their barns. At the end of the year they sell off their tobacco, sugarcane, sesame, etc., and pay off the money lender the principal as well as the interest. Sometimes they sell all their produce to the money lender himself at the market rate. Out of the paddy harvest they return to him one and a half times or one and a quarter times the rice they had borrowed. Whatever they're left with lasts them for the next three or four months. If there's drought in the land or some unusual expenditure and the debtor cannot pay off either the cash or the paddy borrowed, then a fresh account is started for the arrears which accumulate gradually; yet the money lenders never go to court. But as these arrears look like temporary losses they sometimes personally visit the lands to check if the peasant is taking adequate interest in his cultivation; they also verify if he has actually sown the area for what he had borrowed money to pay tax. Occasionally, some short-sighted peasants borrow a lot of money and cheat these money lenders; as a result they always remain in debt and the money lenders suffer big losses and they themselves also are in trouble. To avoid all these difficulties the money lenders go to their fields, but not as 'the blue devils' do. [*Embarassed*]. Your Honour, those Muslim bastards talk like that....

WOOD. You seized by Shani[9]—it force you quit. What reason ferret out these facts otherwise? Why so impertinent? Why? You wicked, incestuous brute!

GOPINATH. Your Honour, we're the ones to be abused, to be kicked, to be sent to jail, but you're the ones to get the applause for opening schools and dispensaries; we're the ones again for all secret murders and kidnappings. If we advise you, you fly into a temper. Only my Gurudev knows, what agony I've been going through because of Majumdar's case.

WOOD. Whenever, ask the sister-fucking swine show guts, bastard quotes Majumdar's case. All along I say you rascal, coward, good-for-nothing. Why not send Nabin Basu Sachiganj lockup and get over?

GOPINATH. You Sir, are my father and mother, to save this poor servant of yours, may I suggest that you may send for Nabin Basu and talk to him about this case?

WOOD. Shut up you bastard of whores, bitch—for you I talk to that dog? You swine, coward, *kayastha*'s son! [*Kicks him, Gopi falls down.*] Before Commission, if sent witness you finish up, you scoundrel, devil's nigger! [*Kicks him again and again.*] You boast your work good like *kaibarta*—you damn swine *kayastha*; tomorrow see work not good, myself send you jail.

Wood and the candidate leave

GOPINATH. [*Gets up and dusts himself.*] Seven hundred vultures are reborn as one single manager of an indigo planter! How else can he digest so many kicks! What kicks, my God!

Voice within: Manager, manager

The slave is coming, Sir. Whose turn is it now? Ah—there are so many cross currents in this ocean of love!

Gopinath Das leaves.

SCENE 2

Nabinmadhab Basu's bedroom—Aduri is weeping, while making the bed

ADURI. Ah, ah, ah, what shall I do, where shall I go? My heart is broken—the way they've beaten him up, he's hardly breathing! *ma thakrun* will die when she sees him. They were so afraid, thinking that he had been kidnapped and taken to the factory, they all rushed outside and stood under a tree there, crying miserably. They didn't notice that he was being carried into the house.

Voice: Aduri, we're bringing him in

Yes, yes, bring him in; there's nobody here.

Enter Sadhu Charan and Torap, carrying the unconscious Nabinmadhab Basu

SADHU CHARAN. [*Placing Nabin on the bed.*] Where's *ma thakrun*?

ADURI. They're outside, under the tree, looking for him. When he [*pointing to Torap*] picked him up and ran, we thought he was being taken to the lockup and they began to cry, as if their hearts would break; I came in to get help. Will mother live when she sees her son dead? You people just wait a little, I'll go and call them.

Aduri leaves. Enter a priest

PRIEST. Oh God! Oh God! Did you have to destroy such a man! How many people now will lose their living! It doesn't look as if *baro babu* will get up again.

SADHU CHARAN. It's God's will. He can bring even the dead back to life.

PRIEST. Bindumadhab has performed the last rites, in accordance with the scriptures, on the third night, by the bank of the Bhagirathi; arrangements for the monthly oblation are being made, because the mistress wants it. It had been decided that after the ceremony they would leave this place, and he had told me that he would never again go and meet those tyrants; then how is it he went to them today?

SADHU CHARAN. It wasn't his fault, nor was he hasty; *ma thakrun* and *bou thakrun* had begged him not to go. They had said that the few days they were still here, they'd bathe in the well water, or if necessary Aduri could fetch some water from the pond; there wouldn't be any trouble. *Baro babu* told them, 'I shall pay *saheb* fifty *rupees* and fall at his feet, beg him not to sow indigo on the sides of the pond and shan't say a word about anything controversial.' Deciding this, he took me and Torap with him and went to the indigo field and told the *saheb*, with tears in his eyes, 'Sir, I've brought fifty *rupees* for you as a mark of obeisance; this year please do not sow indigo there, but if you do not wish to grant this request, please take this money and be merciful to this fatherless, miserable tenant, and stop sowing till my father's funeral is over.' That beast replied—it's a sin even to utter those words, my hair's standing on end, the swine said, 'Thy father had hanged himself like thieves and dacoits do, inside a foreigner's jail; for his funeral several bulls will have to be butchered, keep the money for that.' Then thrusting his boots at *baro babu*'s knee, he said, 'Here are alms for thy father's funeral'.

PRIEST. [*Plugging his ears.*] O God! O my God....

SADHU CHARAN. At that, blood flew into his eyes; he began to tremble with rage, bit his lips and after a moment's silence, he kicked *saheb*'s chest so hard that the brute fell down flat on the ground, like a load of straw. After that Keshe Dhali, who is now the head guard of their factory, and ten other spearmen surrounded *baro babu*. These were the men whom *baro babu* had saved from being convicted of dacoity—the swine felt a little embarassed to strike him—so the big planter got up, and giving a big slap to the head guard he snatched the big stick from his hand and with it he struck *baro babu* on the head. The blow broke his skull and he fell down unconscious. Though I tried hard I couldn't break the circle to reach him, but Torap, who was standing at some distance and saw every thing, rushed in like a mad buffalo the moment he was surrounded, picked him up and ran back.

TORAP. Yes, *baro babu* had told me, 'You stand apart a bit. Who knows, they might grab you again.' Those bastards are furious with me—but would I have hid myself, if I had known that they were going to beat him up! Only if I had reached the spot just a few moments earlier, I could have brought him back alive, and slaughtered those two bastards as sacrifice at Barkat Bibi's shrine. But when I saw his skull, I just lost nerve and couldn't kill those swines. O God! O God! so many times did *baro babu* save me, and I couldn't save him now. [*He weeps, beating his forehead.*]

PRIEST. What's this wound on his chest—was it with a sharp weapon?

SADHU CHARAN. As soon as Torap broke the circle, *chhoto saheb* tried to strike *baro babu*, who had fallen on the ground, with his sword, but Torap warded off the blow with his hand—his left hand has been badly slashed—this wound on the chest is from that sword.

PRIEST. [*Thoughtfully*] Danger is like a touchstone—it tests the true worth of a man's wife, friends and servants. I don't see a soul here, but this man, Torap, from a different village, a different religion, is sitting by *baro babu*, crying; he's a poor man and his hand is his fortune—such a pity—he has lost it. But how could there be such a lot of blood on his mouth?

SADHU CHARAN. When *chhoto saheb* slashed his hand, mad with pain—like a mongoose which bares its teeth and bites when its tail is cut off—Torap bit off *baro saheb*'s nose.

TORAP. Look, I've shoved it down my waist, I'll show it to *baro babu* when he gets better. [*Shows the nose.*] If he hadn't run and saved himself, I'd also have torn off the ears of that son of a bitch—Allah has given him life, so I wouldn't have killed him.

PRIEST. There is justice still—after Surpanakha's[10] nose was cut off, the gods got a respite from Ravana's tyranny; shouldn't the poor peasants also get some relief from the oppression of the planters, now that the big planter has lost his nose!

TORAP. I'll hide myself now inside the grain store; late at night I shall slip out. Loss of his nose will make him let loose hell on the village.

He bows twice on the ground, at the foot of Nabinmadhab's bed and leaves

SADHU CHARAN. *Mathakrun* has become a shadow of herself after Master's death. Now, when she finds *baro babu* like this my fear is she'll die. I've sprinkled such a lot of water, massaged his heart, he's still unconscious, will you call him once?

PRIEST. *Baro babu*! Oh *baro babu*! Nabinmadhab! Nabinmadhab! [*With tears in his eyes.*] Oh protector of the poor! Oh benefactor of the needy! Aren't the eyelids stirring a little? No! No! Mother is sure to kill herself—after the news of the hanging, she vowed to fast for full ten days. Today's the fifth day. Early this morning Nabinmadhab put his arms round her, wept and told her, 'Mother, I know I'll be committing a great sin and go to hell if I disobey you, but I'll go without my ritual rice, and fast if you do not eat something today.' His mother then kissed him and said, 'Son, I was the wife of a king all these days, now I'm the mother of a king. I wouldn't have been so wretched if I could have placed my head on his feet when he died. Such a pure soul, to have such an unnatural death! I'm fasting as an expiation. You're the only treasures of this miserable mother; for the sake of you and Bindumadhab I'll partake of the food offered to our priest. Please don't weep.' Saying this, she took him in her arms, as if he were a five-year old child. [*Wailing within.*] They're coming.

Enter Sabitri, Sairindhri, Saralata, Aduri, Rebati, an aunt of Nabinmadhab, and neighbours

Don't be alarmed, he has still life in him....

SABITRI. [*Noticing the lifeless Nabinmadhab.*] Nabinmadhab, my child, my child, my child where are you, where, where. [*Faints and falls down.*]

SAIRINDHRI. [*Weeping*] Saralata, you mind *thakrun*, let me see my love once, for the last time. [*She sits near Nabinmadhab, looking at his face.*]

PRIEST. [*To Sairindhri.*] *Ma*, you've always been a chaste and loving wife, and you possess all the marks of the auspicious; sometimes a dead husband with such a wife comes back to life. His eyes are still fluttering, nurse him confidently. Sadhu, you be here till Mistress comes round.

Priest leaves

SADHU CHARAN. See if she's breathing—she's so still.

SARALATA. [*Examines if Sabitri is breathing, then tells Rebati softly.*] She's breathing well, but her head feels as if there were fire inside. I can hardly touch it.

SADHU CHARAN. The manager had gone to fetch the *kabiraj*. I wonder if he has fallen into the *saheb*'s hands. I'll go and get him. [*He leaves.*]

SAIRINDHRI. Ah! ah! My love, you were so miserable because she was fasting, and looked after her all the time, seeing that she had gone so thin. For the last few days she couldn't even sleep without having you close to her—and now she's lying by you unconscious. Won't you look at her once? [*Looking at Sabitri.*] Misery, misery—there she is after the loss of her son, the very source of her life, like a dead cow bitten by a snake, when losing her calf she wanders around lowing sadly. Dearest, open your eyes, call this servant of yours, let her hear that sweet voice calling her, and feel happy. The sun of my happiness has set at mid-noon—what's going to happen to my Bipin. [*She falls on Nabinmadhab and cries.*]

SARALATA. Take comfort *didi*.

SAIRINDHRI. [*Sits up.*] I lost my father when I was a child. He had been dragged to their factory for this deadly indigo and he never came back—he met his end in the indigo lockup. My poor mother went with me to my uncle's house—and there she died of grief. Then my uncles brought me up—a dried up, wayside little flower, dropped out of a gardener's flower basket. It was then that my love picked me up with honour and affection; in him I found the parents I had lost, and forgot my loss. Today all those sorrows are coming back afresh. If I lose him, his all embracing love, I shall be reduced to a miserable orphan again. [*She falls to the ground.*]

AUNT. [*Holds her hand.*] Don't be so frightened dear, don't be so upset. We've sent word to Bindumadhab to come down with a doctor, he's sure to cure him.

SAIRINDHRI. Auntie, as a girl I used to wish and pray while lighting the evening lamp; keeping my hand on the rice-powder decoration I used to pray that I might get a husband like Rama, a mother-in-law like Kaushalya, a father-in-law like Dasharath and a brother-in-law like Lakshman,[11] and Auntie, God granted my prayers beyond my dreams. I had a husband, extraordinarily handsome, a protector of his peasants, a mother-in-law who showered unceasing love, the very image of sweetness, a father-in-law always cheerful, always affectionate—just happy to call me his daughter-in-law—and almost effulgent in his beauty, and a brother-in-law purer than the radiant light of the autumn moon, dearer to me than Lakshman ever was to Sita Debi. I had been given every thing—but now I find that there's something I haven't got. How's it that I'm still alive—my Rama's on his way to the forest, and no arrangement has been made for Sita to accompany him? Ah, ah, how wretched he was because father had died fasting, and now before the rites are over he himself is gone. [*Gazing at his face.*] Oh such misery, the lips of my love are so parched—please fetch my Bipin from school, let him pour a few drops of Ganga water into his father's dry mouth.[12]

ALL OTHERS. Ah! Ah! Ah!

AUNT. [*Lifting her up.*] My child, don't say such things. [*She weeps.*] If sister had been conscious, these words would have killed her.

SAIRINDHRI. Ma, my husband had to suffer a lot in this world. I want him to be happy in the other world. Dearest, this servant of yours will pray to God as long as she lives—my love, all your life you've been deeply religious, always helped others and protected the poor. That great Friend of the helpless—Bisheshar[13]—will accept you as His. Oh my love, oh my love, take this servant with you; she'll pluck flowers when you go to worship the gods. What misery is this! Rama's leaving for the forest without his Sita! What shall I do, how shall I live? Oh God, O Friend in my distress, come to my help! Oh Protector of women, come to her rescue! Indigo fire is consuming Nabinmadhab. Where are You, Lord of the poor? The lord of my life is leaving me and I'm so miserable. [*She feels Nabinmadhab's heart and sighs.*] Leaving all his loved ones at Your feet, he's taking shelter with You; You are the Ocean of mercy, Refuge of the sinner; grant him, the life of my life, salvation.

SARALATA. *Didi*, *thakrun* has opened her eyes, but she's looking at me so fiercely. [*Weeps*] *Didi*, I've never seen *thakrun* looking at me so angrily.

SAIRINDHRI. *Thakrun* loves Saralata so much—because she's not in her senses, this angry look of hers makes this little flower of ours feel that she has been thrown into a frying pan with hot sand. Sister don't cry; as soon as she's fully conscious she'll kiss you again and tease you as the mad woman's daughter.

SABITRI. [*She gets up, sits by Nabinmadhab and gazes at him cheerfully.*] There's no pain like the labour pain, but when I look at this priceless treasure, I forget all pain. [*Crying*] Ah misery, if only the *bibi* hadn't sent a letter to the God of Death to kill my husband! How happy he'd have been to see this lovely child. [*She claps her hands.*]

ALL OTHERS. Ah! Ah! Ah! She's gone mad!

SABITRI. [*To Sairindhri.*] Oh midwife, give my baby to me, my burning body will feel cool then; let me kiss him in the name of his father. [*Kisses Nabinmadhab.*]

SAIRINDHRI. Mother, I'm your *baro bou*, can't you see me? Your life's treasure is lying there unconscious, he can't talk.

SABITRI. He'll talk when we have his first rice-eating ceremony. If the Master had been here today there'd have been so much merriment, so much singing and playing of music. [*Cries*]

SAIRINDHRI. Misfortune upon misfortune—*thakrun* has gone mad.

SARALATA. *Didi*, take Mother away from the bed. I'll nurse her, she might feel better.

SABITRI. Should you have written such a letter! Such a happy day today and we can't have any music? [*She suddenly gets up, looks around and goes to Saralata.*] I beg you, *bibi thakrun* please, please write another letter so that God of Death might send the Master back; if you hadn't been *saheb*'s *bibi*, I'd have held you by your feet.

SARALATA. Oh Mother, you've always loved me more than my own mother. Mother, these words from you are more hurtful than any torture of death. [*She clasps Sabitri in her arms.*] Mother, to see you like this, makes me feel as if there were a fountain of fire inside me.

SABITRI. You whore, you evil creature, you disgusting foreign woman! How dare you touch me on the *ekadashi*[14] day? [*Frees herself.*]

SARALATA. Oh, Mother, Mother, after these words I don't want to live any more. [*Holds Sabitri's feet and falls down, crying.*] Mother, it's better that I die at your feet.

SABITRI. Good thing, the harlot woman's dead, my Master has gone to heaven, but hell is for them, wicked creature! [*She laughs and claps her hands.*]

SAIRINDHRI. [*Getting up.*] Ah, my poor, gentle Saralata, and such a beloved *bou* of my mother-in-law, how hurt she is. [*To Sabitri.*] Mother, you come to me.

SABITRI. Midwife, you left the baby alone dear? Let me go to him.

She runs and sits by Nabinmadhab's body

REBATI. Mother, Mother, don't you remember how you used to say, 'In this whole village no one has a daughter-in-law like my *chhoto bou.*' You never ate before feeding her—and now you're abusing her as a harlot? Mother, can you hear me? It's you who brought us up, don't you remember?

SABITRI. Yes, yes, come when my baby's eight days old, you'll eat here that day.

AUNT. *Didi*, your Nabin will come round, don't lose your mind.

SABITRI. Now, how did you know—nobody knows this name! My father-in-law used to say, 'When *bouma* gets a son, I'll call him Nabinmadhab.' Now that I've got a son, I'm going to give him that name. The Master used to say, 'When are we going to get a son that I might call him Nabinmadhab?' Only if he'd been alive today all his wishes would've come true. [*She cries.*]

[*Noises within.*] There, there, the musicians are coming. [*Claps her hands.*]

SAIRINDHRI. The *kabiraj* has come, *chhoto bou* please go to the other room.

Enter the kabiraj *and Sadhu Charan; Rebati, Saralata and neighbours leave the room. Sairindhri stands on one side, her* sari *pulled down over her head.*

SADHU CHARAN. Thank God, *Ma thakrun* is sitting up.

SABITRI. [*Crying*] Only because my Master isn't here, you had to leave the drum behind, and on such a day as this?

ADURI. Is she in her senses now? She has gone stark mad! She's calling *baro babu* her infant son—and she abused *chhoto bou* so badly as *'bibi'* and what not! Poor *chhoto bou* has been crying her heart out. And now she's calling you people musicians!

SADHU CHARAN. Ah, how dreadful!

KABIRAJ. [*Sitting by Nabinmadhab*.] Well, grieving for her husband she had been fasting, and then this sudden shock of seeing her beloved son in this condition. There can be sudden madness in such cases and it's according to medical science too—I'll have to feel her pulse. Mother, please show me your hand. [*Stretches his hand towards her*].

SABITRI. You son of a whore—you must be from the *saheb*'s factory—why else should you like to hold a good woman's hand? [*She gets up*.] Midwife, look after the baby, I'll go and have a drink of water. I'll give you a nice silk *sari*. [*She leaves*.]

KABIRAJ. What a pity! That lamp which illumines our mind will never again shine; I shall send some Himsagar oil; that's what's used these days. [*He feels Nabinmadhab's pulse*.] The pulse is extremely weak, I can't find any other symptom. Though Western-type doctors are otherwise useless, they know surgery—it will be expensive, but I think you should send for one.

SADHU CHARAN. *Chhoto babu* has been asked to come down with a doctor.

KABIRAJ. Good.

Enter four relatives of the family

FIRST RELATIVE. Who could have ever dreamt of such a thing happening! And that too at noon! Some were sitting down to eat, some were having bath, some were resting after a meal—I heard about it only now.

SECOND RELATIVE. It's so dreadful! The wound in the head looks very serious. There was no possibility of a showdown today, otherwise all the peasants would have been there.

SADHU CHARAN. Two hundred peasants carrying sticks are roaming around in a fighting mood, they're all crying for *baro babu*, calling him—I've asked them to go home. On the slightest provocation he'll burn down the village, and now that *saheb* has lost his nose!

KABIRAJ. Wash his head and pour a little turpentine oil on it. Later in the evening, I'll come again and prescribe something else. See that there's no noise in the patient's room; it increases the illness—nobody should talk here.

The kabiraj, *Sadhu Charan and relatives leave. Aduri goes inside and Sairindhri sits down*

SCENE 3

A room in Sadhu Charan Ghosh's house: Kshetramani lying in bed in great pain. By her bedside are Rebati and Sadhu Charan

KSHETRAMANI. Smooth out the bed, Mother, smooth out the bed.

REBATI. My darling, my sweet lovely child, what's troubling you dearest? I've smoothed out the bed, there's nothing on it my love, I've spread on our cloth the quilt your aunts sent from the Master's house.

KSHETRAMANI. Thorns are pricking me all over—they're killing me. Mother, turn me towards Father.

SADHU CHARAN. [*Gently turns her. Aside*.] Thorns in bed—not long now—symptoms of approaching death. [*Aloud*] My *ma*! my treasure! Eat something! I got some pomegranates for you, from Indrabad. I bought for you your favourite lime-coloured *sari*. You'd be so happy to see it, but you haven't even seen it.

REBATI. Yes, I knew my darling's wishes—she had specially asked for a necklace of shells for her sixth month celebration. Ah, ah, my beautiful child's looking so different! What am I to do? God, God. [*Flings herself on Kshetramani.*] My golden girl, my golden girl! Look, look, she's turning dark in the face, look at her eye-balls.

SADHU CHARAN. Kshetra, Kshetra, look at me. Open your eyes, my dearest.

KSHETRAMANI. Axe is cutting me Father, Father. [*She turns on her side and groans.*]

REBATI. I'll put her on my lap. My baby will be more comfortable that way. [*She goes forward to lift her.*]

SADHU CHARAN. Don't lift her, she might faint.

REBATI. My evil luck, my destiny! Ah! Ah! And Haran was such a handsome lad—the very image of the god Kartik on his peacock! Oh God! oh God!

SADHU CHARAN. Ray is taking such a long time over it!

REBATI. *Baro babu* brought her back to me from the tiger's jaw, but that son of a whore had punched her belly so hard that it brought on the abortion, and now it's a struggle for life. Ah, ah, it was a grandson—covered in blood, but perfect in form even to the finger tips. *Chhoto saheb* destroyed my Kshetra, and *baro saheb* finished our *baro babu*—ah! ah! ah! There's none to save the poor.

SADHU CHARAN. Did we deserve a grandson? Were our deeds good enough for that?

KSHETRAMANI. Mother, Mother! I'm being cut to pieces! Oh my waist—sharp *tengra* fish—oh! oh! oh!

REBATI. The night of the *nabami*[15] is over and my golden Durga is going to be immersed in water. Now how shall I live? Who'll call me '*Ma*'? Why, oh, why did we bring her down? [*She falls on Sadhu Charan's shoulders and cries.*]

SADHU CHARAN. Be calm, don't cry, she might faint.

Enter Ray Charan with the kabiraj

KABIRAJ. What are the fresh symptoms? Has she had the medicines I had prescribed?

SADHU CHARAN. She couldn't retain it—she threw up whatever she had taken earlier. Please feel her pulse, it looks as if the end is near—all the symptoms are there.

REBATI. All the time she feels something's pricking her. I've made the bed so soft, yet she's restless; please give her some other good medicine to save her life! Ah! we used to be so happy with our new relatives. [*She weeps.*]

SADHU CHARAN. I can't find the pulse now.

KABIRAJ. [*Feeling the pulse.*] In such cases it is better that the pulse should be weak. We say, 'If the pulse is rapid when the patient is weak, it's sure to kill him.'

SADHU CHARAN. I don't know if any medicine can help now, at best it can give us a little hope; see what you can do.

KABIRAJ. A full dose of Suchikabharan is usually given. Some water in which grains of unboiled rice have been soaked is necessary.

SADHU CHARAN. Ray Charan, in that room there's some rice *thakurani* had sent us for the *swastayan*[16] ceremony, get it quickly.

Exit Ray Charan

REBATI. Ah, if only our *annapurna ma* had been in her senses, she would have come to see my Kshetramani, and brought the rice herself. It's my great misfortune that she has gone mad.

KABIRAJ. Yes, she had already been stricken by the death of her husband, then she saw her son half-dead. Her madness is getting worse, and she's looking so very thin and weak. Maybe she'll pass away even before Nabin.

SADHU CHARAN. How's *bdro babu* today? I somehow feel that with his own sacred blood he has ended the intolerable oppression of those beasts—those planters. The Commission might be of some help to the peasant, but what's the use now? I can stand anything—the bites of a hundred cobras of the Chetan swamp, a sudden fall into the treacle syrup boiling in a huge cauldron over a brick oven's woodfire. I can even stand dacoits piercing my eyes on a new moon night, after they had killed my only son—gentle and educated—kicked his pregnant wife—beautiful and loving—so that she miscarried, and then fled with all the wealth earned over the years by seven generations! All this I'm prepared to suffer; let there be ten indigo factories instead of one right inside our village; but the loss, even for a moment, of *baro babu*—the protector of the peasants, I'm finding impossible to bear....

KABIRAJ. The blow has seriously damaged the skull—the brain has come out; disorder of the three humours has set in and the end will come either at noon or in the evening. Bipin poured a few drops of Ganga water between the lips, but they didn't go down. Nabin's wife, though beside herself with grief, is anxious that all the rites be observed for his salvation.

SADHU CHARAN. Ah, if the mistress hadn't lost her senses she'd have died of a broken heart. The doctor also thinks that the wound in the head is very serious.

KABIRAJ. The doctor has been very kind. When Bindu Babu wanted to pay his fees, he said, 'You're in so much trouble that it may not be possible for you to meet all the expenses for the funeral rites of your father—I really can't take anything from you now. The palanquin bearers who brought me here will take me back—you don't have to pay them anything.' If it had been Doctor Duhshashan, he'd have taken the money kept for the Master's funeral. I've seen that fellow twice—he's so rude and so devilishly greedy.

SADHU CHARAN. *Chhoto babu* had brought the doctor here too, to examine Kshetramani, but he didn't prescribe any medicine. Observing that we'd nothing to eat because of the planter's oppression, he gave me two *rupees* in the name of Kshetramani.

KABIRAJ. Doctor Duhshashan wouldn't even have felt her pulse, but would have told you, 'She won't live', and would have made you sell your cows to pay his fees.

REBATI. If only someone would save my Kshetra. I'll sell off everything to pay him.

Enter Ray Charan, with rice

KABIRAJ. Wash the rice in a stone vessel, and bring me that water.

Rebati washes the rice

Don't put too much water. It's a nice bowl!

REBATI. The mistress had gone to Gaya once, and had brought back a number of such bowls. She gave one of them to Kshetra. Ah, that kind mother is insane today, and to prevent her from hurting herself, they've tied her hands up.

KABIRAJ. Sadhu, get a mortar and pestle. I'm taking out the medicine. [*Opens his box*.]

SADHU CHARAN. Sir, don't take the trouble—see her eyes—Ray Charan, come here.

REBATI. Oh God! Oh God! Is this what fate had decreed? To lose Haran also—that lovely boy! Oh Kshetra, Kshetra, my Kshetramani, my darling. Won't you talk to me ever again, my treasure—ah! ah! ah! [*She moans*.]

KABIRAJ. It's all over.

SADHU CHARAN. Ray Charan, hold the end of the bed.

Sadhu Charan and Ray Charan carry out the bed, with Kshetra's body lying on it

REBATI. [*Running after them*.] I can't, I can't let my golden Lakshmi go away—Kshetra, my Kshetra where shall I go, how shall I live? Even staying with *saheb* would have been better—I wouldn't have lost her at least, I could have seen that sweet face and felt peaceful. Oh! oh! Kshetra, my Kshetra!

KABIRAJ. Ah, such pain, such unbearable anguish of a mother—it's better not to bear children.

SCENE 4

Inner veranda of Golak Chandra Basu's house—Sabitri is sitting with Nabinmadhab's body on her lap

SABITRI. Come, sleep. Come, come my little treasure, my heart's jewel, my little son. Whenever I see this beautiful face, I remember that other face. [*She kisses him.*] My little one is fast asleep. [*She strokes his head softly.*] Ah, ah, what awful mosquito bites—it's hot, but we've to put up the mosquito net. [*Touching his chest.*] Dear, dear, how much can a mother tolerate! Those bed bugs have been at him—such a lot of blood is oozing out of this soft body. Nobody has made his bed yet—where am I to lay my little darling now? I've nobody—none to call my own—lost everyone when I lost the Master. [*She cries.*] Now, look at me, this wretched me, crying when my baby's on my lap! [*She looks at his face and kisses him.*] My treasure is smiling in his dream—no, no my darling, I'm not crying any more, your face has made me forget all my sorrows. [*She gives him her breast.*] Suck me, my little one, drink your milk. How much I begged that harlot, I fell at her feet, but she wouldn't bring the Master back—he could have come at least once, arranged for my baby's milk and then gone back! That woman and the God of Death are so close—one letter from her, and he'd have released the Master. For the husband's salvation, his widow must never wear bangles—how much I screamed and cried, but they had to put these conch bangles on my wrist—I tried to burn them down on the flame of the lamp, but they're still there! [*She bites off the cord.*] A widow doesn't look nice with ornaments on her, nor is it proper. My hands have such blisters on them! [*She cries.*] Within three nights [*Cracks her fingers on the ground*]—may she, who has made me a widow, become a widow herself, having no right to wear conch bangles ever again! Never mind, I myself will make my baby's bed. [*She acts as if making the bed.*] Oh, the mat hasn't been washed for days. [*Stretches her hand out.*] I can't reach the pillow and the quilt is so soiled. [*She wipes the floor with her hands.*] Now let me put my baby down. [*She gently lowers Nabin's body on to the floor.*] Your Mother's with you, so why should you be afraid? Sleep well, my treasure. Let me spit to ward off the evil eye. [*She spits on his chest.*] If that whore, that *bibi*, comes again today, I'll strangle her. I shan't leave my baby alone today for a moment. Let me draw the line of charm.[17] [*Muttering, she makes a line with her finger on the floor around Nabinmadhab's body.*]

Cobra's froth and tiger's nose,
Burning incense and Charokh[18] *swinging,*
Grey hair of seven co-wives,
Bhanti[19] *leaves and flowers*
Of the horsechestnut,
Indigo seeds and burnt chillies.
Dead man's skull and Madar's[20] *root,*
Famished dogs and
Chandi *the goddess of thieves:*
This is the line which Yama *cannot bite.*

Enter Saralata

SARALATA. Where have they all gone? Ah, here she's—going round the body. My man must have been very tired after the journey; on the floor itself he has fallen into a deep sleep which makes one forget one's sorrows. Sleep can work such wonders—a widow finds her man back, a home-sick stranger finds himself back there, prisoners with shackles become free once again. It's the best medicine for the sick, it's law is meant for all—regardless of caste. My man is in the kingdom of sleep, free of all his worries. And that explains how Mother could carry the body of her dead son here, lying by his side. He's in deep anguish since he lost his father and brother, his face has lost its brightness, just as the moon does when the dark fortnight sets in. Oh Mother, when did you get up and come here? I've been with you all the time—never bothered to eat or sleep—could it be that I too had fallen asleep? Mother, I've promised to get Father back from King Yama; that made you feel a little better? The night is deep, it's so dark as if the end of creation is nigh, black clouds have gathered in the sky and flashes of lightning look like streaks of fire. All are asleep, how silent it is—only the jackals howl in the depth of the woods and the dogs bark. How could you come out alone, Mother, in this dreadful night, carrying the body of your dead son? [*She approaches Nabinmadhab's body.*]

SABITRI. I drew the charmed line and you crossed it?[21]

SARALATA. Ah, ah, what a wonderful brother he was! Will my husband be able to bear this terrible loss! [*She weeps.*]

SABITRI. Oh, you evil woman! You bitch! You whore! You're jealous of my baby? May your husband die! May you become a widow! Get out from here, get out—I'll throttle you if you don't.

SARALATA. Such a pity, such a pity, to have lost such a son!

SABITRI. I forbid you, you harlot, to look at my baby! You're tempting death, I warn you!

SARALATA. [*Approaching*] Oh Yama, how cruel you are to make this simple, kind mother-in-law of mine suffer so much! Yama! You are so heartless!

SABITRI. So you are calling him again? Again and again? [*She seizes Saralata by the throat and throws her to the ground.*] You wicked bitch, Yama's mistress, come, let me finish you off! [*She stands on Saralata's neck.*] You've eaten my husband and now you're calling out to your lover to eat up my baby! Die! die! die! [*She dances on her throat.*]

SARALATA. Ah...ah...ah....[*She dies.*]

Enter Bindumadhab

BINDUMADHAB. Oh, you're here! Good God! Mother, Mother, you've killed my Saralata! My sweet love has left this wretched world. [*He weeps and kisses Saralata.*]

SABITRI. Bite her, bite her, kill that evil woman! She was calling Yama to eat up my little one, I had to kill her, standing on her throat.

BINDUMADHAB. Oh Mother, my Mother, if by sheer chance at night, a mother, while suckling her infant, crushes it to death in her sleep—she kills herself when she wakes up in the morning and finds what she has done. I fear that you'll do the same if this madness which makes one forget one's sorrow leaves you, and you realize what you did to your beloved Saralata. Perhaps it's better you don't get your mind back again. It's such a comfort to a woman grieving over the loss of a husband and a son; madness is like a protective wall against the deadly attack of grief. Mother, I'm your Bindumadhab.

SABITRI. What? What did you say?

BINDUMADHAB. Mother, I can't bear this life any more! Mother, Mother—Father's suicide, brother's death, your madness—and now in your madness you've killed my Saralata!

SABITRI. What? My Nabin is dead? My Nabin is gone? Oh, oh my treasure, my Bindumadhab I have killed my Saralata? My *chhoto bou-ma*! I killed? I went mad and killed her? [*She rushes to Saralata and embraces her.*] Ah, ah, I might have lived on even after losing my husband and son, but the sorrow of killing you is too much for me to bear—oh! oh! *ma*! [*She dies.*]

BINDUMADHAB. [*Feels Sabitri.*] Just as I feared! The moment she regained her senses she died! Now there's no mother to love or kiss me—nobody to call '*ma*'. [*He weeps.*] Let me touch your feet for the last time—let me taste the dust of these feet and feel blessed.

Enter Sairindhri

SAIRINDHRI. Little brother-in-law, I've decided to become a *sati*.[22] Please don't try to stop me. Saralata will be a mother to Bipin, but what's this? Why are these two lying like this? Oh! oh!

BINDUMADHAB. *Baro bou*, Mother killed Saralata in a fit of madness—then suddenly she got back her mind and the violence of her grief killed her.

SAIRINDHRI. Now? Just now? How terrible! What to do now, what to do! My sweet sister, that new cord for your hair—you never used it even once. Ah, ah, who'll call me *didi* again? [*She cries.*] And *thakrun*, you've gone to your Rama, and you didn't let me go! Mother, Mother, not for a day did I feel that I had no mother, you were like my own mother. [*She cries again.*]

Enter Aduri

ADURI. Come quickly, Bipin has woken up in a fright.

SAIRINDHRI. You shouldn't have left him alone. [*She rushes out.*]

BINDUMADHAB. My Bipin will be my only pole star in this turbulent ocean. Man comes into this world for a while to play his part—as ephemeral as the high banks of a swift river. Yet how beautiful is the bank of the river—covered with soft green *darba* grass, with mighty trees bedecked with fresh leaves standing on its borders, where cattle graze with their calves, peacefully munching the luscious grass. A strole there, the sweet songs of the birds ringing in one's ears, the flower-scented breeze fanning one—all these fill one with the thought of that Joyous Eternal Being. Then, all of a sudden, there appears a thin line on the field, and the bank with all its beauty tumbles down and disappears deep inside the swirling waters. What misery! The entire Basu family has been swallowed up by this Kirtinasha,[23] this Indigo. Oh Indigo, what a killer you are!

EPILOGUE

The Indigo Planter is a venomous snake,
My happiness he has reduced to ashes.
Injustice killed father in the prison,
Brother was felled in the indigo field,
Grievous loss made mother mad, and
In a fit of madness lovely Saralata she killed.

My anguished cries brought her senses back,
That very moment rose wave after wave of grief
In the mighty ocean of sorrow.
And pain, like a sharp-edged sword, pierced her.
Inconsolable, in great anguish, my mother died.

Father, oh Father, I call you so often,
Won't you come back smiling
And press me to your heart?
I look for you all around, Mother, oh Mother,
But nowhere do I find that happy, cheerful face.
Ah, the depth of a mother's love for her son,
The wonder of that love—immeasurable, limitless.

Brother is gone too—a greater friend had no man,
Loving, affectionate, life's precious gift.
Open your eyes once, brother—Bindu has come home.

Oh, my heart is breaking.
Where have you hidden yourself, my sweet, sweet love
My beautiful, accomplished, loving wife—
My doe-eyed beauty, with the gait of a swan!
How you used to read to me the stories of Betal,[1]
My hand in yours—casting a spell
On my engrossed mind.
That music of your voice, sweet and enchanting,
As if in a deep wood a bird were singing.
She was a lotus of light and beauty,

Floating in my heart and making it radiant.
Who stole my lotus—whose cruel hand?
Leaving me desolate in this eternal darkness!

This world today is a cremation ground,
Strewn with corpses of those I love.
Gone is my father, my mother, my brother;
Gone is my wife—I am left alone.

Ah—where have all those people gone
In search of my brother's body?
When they come back the funeral must be
Prepared on the banks of the Jahnavi.
Ah—such a wonderful man Nabinmadhab was,
A lion among men—what an end to what a life.

[He sits down, holding his mother Sabitri's feet]

Here ends the drama *Neel Darpan.*

NOTES

Prologue

1. Name of a circular weapon used by the God Krishna and mentioned in the epic Mahabharata.

Act I.

1. Perfect as in the 'Golden Age'.
2. 'Him' refers to her husband, whose name she is not allowed to use in keeping with Hindu tradition.
3. A play on words; the Bengali term *dan* can mean, either 'right', or 'witch'.
4. This is a reference to a literary work by Ishwar Chandra Vidyasagar.
5. Vermilion applied to the parting of the hair is the sign of a married woman.
6. A reference to the Mother Goddess in her aspect as protector of children.
7. The term 'mother' (*ma*) is not only a kinship term; it is an expression of affection and tenderness towards younger women, and a general term of respect towards elder women.
8. In traditional Hindu India it is not considered proper for young husbands and wives to show their mutual feelings of affection openly.
9. One of God's names, invoked here to express both horror and pity.
10. Onions are classified as 'hot' in the Hindu humoral system, and hence not eaten by many. Widows, such as Aduri, may not eat them.
11. Beards were not common among Hindus.
12. A native of Arakans, Burma, known as violent and lawless; here a lawless country. A corruption of 'Magh', the term by which Buddhist communities of northern Bengal (now Bangladesh) are known. Magh derives from the ancient Indian Buddhist centre, Magadha.
13. Litt. 'head of household', i.e. her husband, Golak Chandra Basu.
14. The 'outcastes', the riff-raff.
15. Corruption of the word 'Magistrate'.
16. An aspect of the Mother Goddess.
17. Corruption of the English word 'appeal'.

18. Litt. the Goddess presiding over a kingdom's welfare.
19. See note 7.

Act II

1. Gajendra, the name of a celestial being.
2. Caste restrictions did not allow a Muslim to stand on a Hindu's shoulders.
3. Reformist Hindu movement founded in the nineteenth century by Raja Rammohan Roy. The movement was strictly monotheistic and focussed among others on the emancipation of women from various Hindu traditions, such as child marriage and widow burning (*sati*).
4. Yama is the God of death.
5. An aspect of the Mother Goddess.

Act III

1. Bhima and Duhshashana are two characters in the epic Mahabharata. Princess Draupadi was forced to give in to Duhshashana's lust, but was saved by Bhima, who was known for his courage and strength.
2. *Karta*: litt. the worker, the doer; i.e. the head of household.
3. Name of the Mother Goddess.
4. The last journey to the Ganga symbolizes for every Hindu death and the last rites.

Act IV

1. This is a reference to the traditional Indian medical concept of hot and cold humoral states.
2. This is a reference to the necktie, which was introduced by the Europeans.

Act V

1. The sacred thread is the prerogative of the Brahmins.
2. Litt. 'filled with rice', an epithet of the Mother Goddess; also used to describe a woman who is generous.
3. Litt. 'Oh, the frog has caught a cold!'.
4. Nanda is a name of the clan of Krishna, the cowherd-God who is depicted in mythology as a great teaser.
5. The female deity who presides over chicken pox and small pox.
6. Name of a low Hindu caste.

7. The Indigo Commission.
8. Queen Sumitra was the mother of Lakshmana and Shatrughna, two of the heroes of the epic Ramayana. Lakhsmana is said to have been killed by a powerful missile, the Shaktishel, thrown by Meghanada, the son of Ravana, the demon king of Lanka.
9. Shani = Saturn; a reference to the destructive qualities imputed to this planet and the ruin which comes upon all those who are under its influence.
10. In the epic Ramayana, Surpanakha and Ravana are the evil siblings against whom Rama and his army emerge victorious. Lakshmana, the devoted brother of Rama cut off the beautiful Surpanakha's nose, when she tried to seduce Rama and sprang on his wife Sita in an attempt to kill her.
11. Principal characters of the Ramayana who symbolize the good.
12. The pouring of Ganga water into the mouth of a dying man by his son is highly coveted by Hindus.
13. The Lord of the Universe, an epithet of God.
14. The 11th night of the fortnight and a day to which special astrological significance is attached.
15. This is the 9th night of the bright fortnight in the month of Ashvina when, at the end of the Durga Puja, Durga's idol is immersed in water as a symbol of her annual departure from her parents' home to that of her husband.
16. A religious service performed, among other things, for the recovery from illness and to ward off evil after death in a family.
17. This reference to an act of magic is drawn from the Ramayana, where Lakshmana draws a protective line around Sita, when she is left alone in their forest dwelling, while the men go hunting.
18. Ceremony performed on the last day of the month of Chaitra in honour of Shiva. It consists of swinging from a tall bamboo pole and thereby piercing various parts of the body.
19. Name of an ornamental plant (*Volkameria infortunata*).
20. Coral tree known to be very thorny.
21. See note 17.
22. A *sati* is a widow who immolates herself on her husband's funeral pyre.
23. An epithet of Yama, the God of Death.

Epilogue

1. A reference to the collection of Sanskrit stories, the *Betal Panchavimshati*

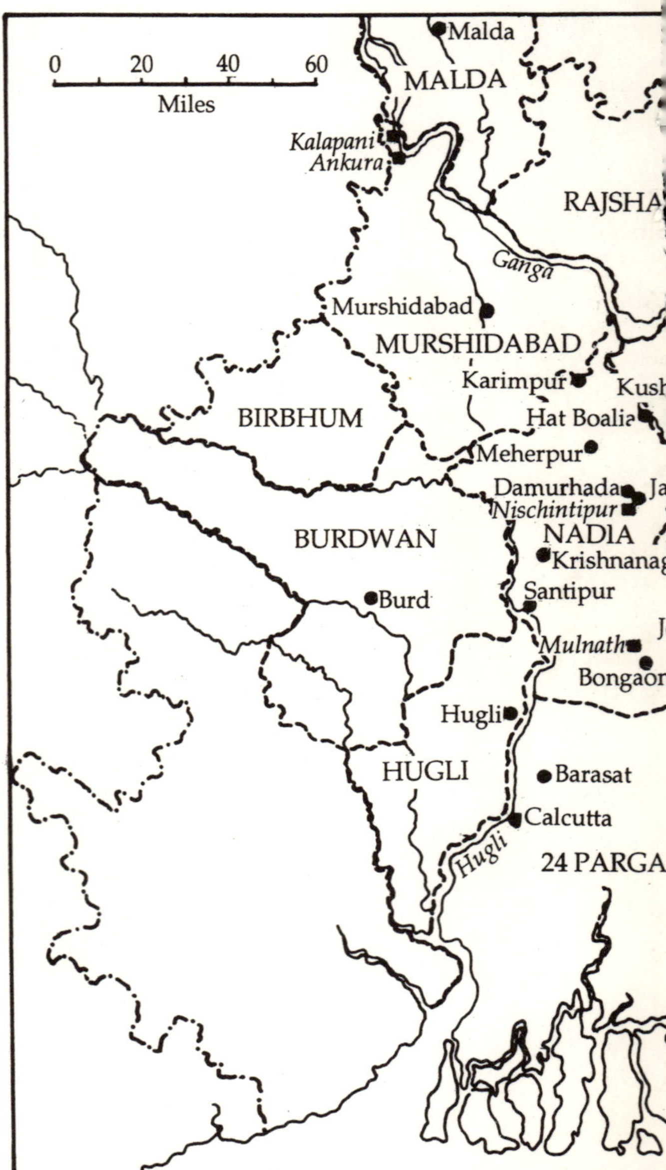

0
20
40
60
Miles
Malda
MALDA
Kalapani
Ankura
RAJSHA
Ganga
Murshidabad
MURSHIDABAD
Karimpur
Kush
BIRBHUM
Hat Boalia
Meherpur
Damurhada
Ja
Nischintipur
BURDWAN
NADIA
Krishnanag
Burd
Santipur
Mulnath
Bongaon
Hugli
HUGLI
Barasat
Calcutta
Hugli
24 PARGA